Beekeeping
A Seasonal Guide

Beekeeping
A Seasonal Guide

Ron Brown

B. T. Batsford Ltd · London

FRONTISPIECE Drone and workers on comb
(photo: G. F. H. Lawes).

First published 1985
Reprinted 1989
Reprinted with corrections 1992

ISBN 0 7134 4489 4

Typeset by Keyspools Ltd, Golborne, Lancs
and printed in Great Britain by
Courier International Ltd
East Kilbride, Scotland
for the publishers
B. T. Batsford Ltd
4 Fitzhardinge Street
London W1H 0AH

Contents

Preface and acknowledgements

Many beekeepers have asked me, 'What book is there which explains what we should be doing at the various seasons of the year, stage by stage?' This volume is intended to fill that need, and at the same time to provide material at three different levels: 'Beginner', 'Improver' and 'Experienced'. The hoped-for progression is that one should be a beginner for two years, an improver for the next three and after that be regarded as experienced. This is not necessarily linked with owning large numbers of hives, and there will be many beekeepers who remain content with two hives and one or two nuclei, yet are still able to experiment with various techniques of management and queen-rearing.

Much thought has been given to the question of metric versus imperial units. Some books quote both throughout, which can be infinitely tedious. In view of the probability that readers of this book are likely to be much happier with inches, pounds and degrees Fahrenheit, it has been decided to use these units and rely on the comprehensive conversion tables given at the end of the book, with just an occasional use of both where it might be helpful. This decision is reinforced by the fact that modern American bee books also use the old, familiar units, and the attitude of 200 million readers of English is important.

Where dates, months and seasons have been quoted, it is with the northern hemisphere in mind. Readers in southern hemisphere countries will appreciate the reversal of our seasons.

Some of the material in this book has already appeared in articles written by me and published in *Beekeeping* and *Home Farm*, periodicals listed in Appendix B. The map on page 28 is adapted from *The Gardening Year*, © 1982, The Reader's Digest Association Ltd, London, and is used with permission.

I wish to express my gratitude to all those who have helped in the production of this book: to my publishers for their patient help and encouragement, to beekeeping friends all over the world (especially in Central Africa, Australia and New Zealand), but above all to my dear wife Rose who typed the entire first draft from my pencilled notes, as well as most of the final manuscript, and encouraged me at every step.

One

Beginning with bees

WHY KEEP THEM?

Most people say 'To get honey, of course,' and they are right, but there is much more to it than that. Firstly, the honey that you will get will be far, far better than most of the honey you can buy. If you choose to eat your own honey, unstrained and unheated (or in the comb) it will contain pollen and propolis, and be a most valuable health food; indeed, a book published a few years ago established a connection between the intake of pollen and healthy longevity in five different parts of the world where unusual numbers of centenarians are found (*The Secret of Staying Young*, Lyall and Chapman).

A fair estimate of the amount of honey which you might expect in a poor-to-average district, year in and year out, would be about 20–40 lb per hive, so that two hives would keep a family in honey all the year round, with a surplus for Christmas presents, swaps with neighbours who have spare apples and vegetables, and so on. Then there is beeswax as a by-product, not much in your first year but afterwards perhaps 1–3 lb annually from your two hives; enough to make all the cold cream, face cream, lipstick and furniture polish your family will need. If you want to go in for candle-making then you must have more hives, or buy in wax from beekeeping colleagues.

Your fruit trees (and those of your neighbours) will yield more heavily, as will beans, gooseberries, raspberries, etc., but above all you will be launched into the most fascinating hobby and will make lots of new and interesting friends. In our bee clubs we have doctors and chimney-sweeps, company directors and lorry-drivers, policemen and postmen, carpenters and teachers, housewives and parsons. We have social outings in summer, and dinners and games in winter, quite apart from bee talks and demonstrations.

The snags

Don't they sting people? Well yes, they can sting, but unlike wasps they die when they do, and so normally they go happily about their business. For years I have had at least a dozen hives in my garden in the middle of Torquay and am still on excellent terms with my neighbours. The strange thing is that if bees see moving humans most of the time, they seem to learn to accept them. I often work around my hives in summer wearing only shorts and shoes, resting seed trays on hive roofs and forks or rakes against hives, but would not dare to do

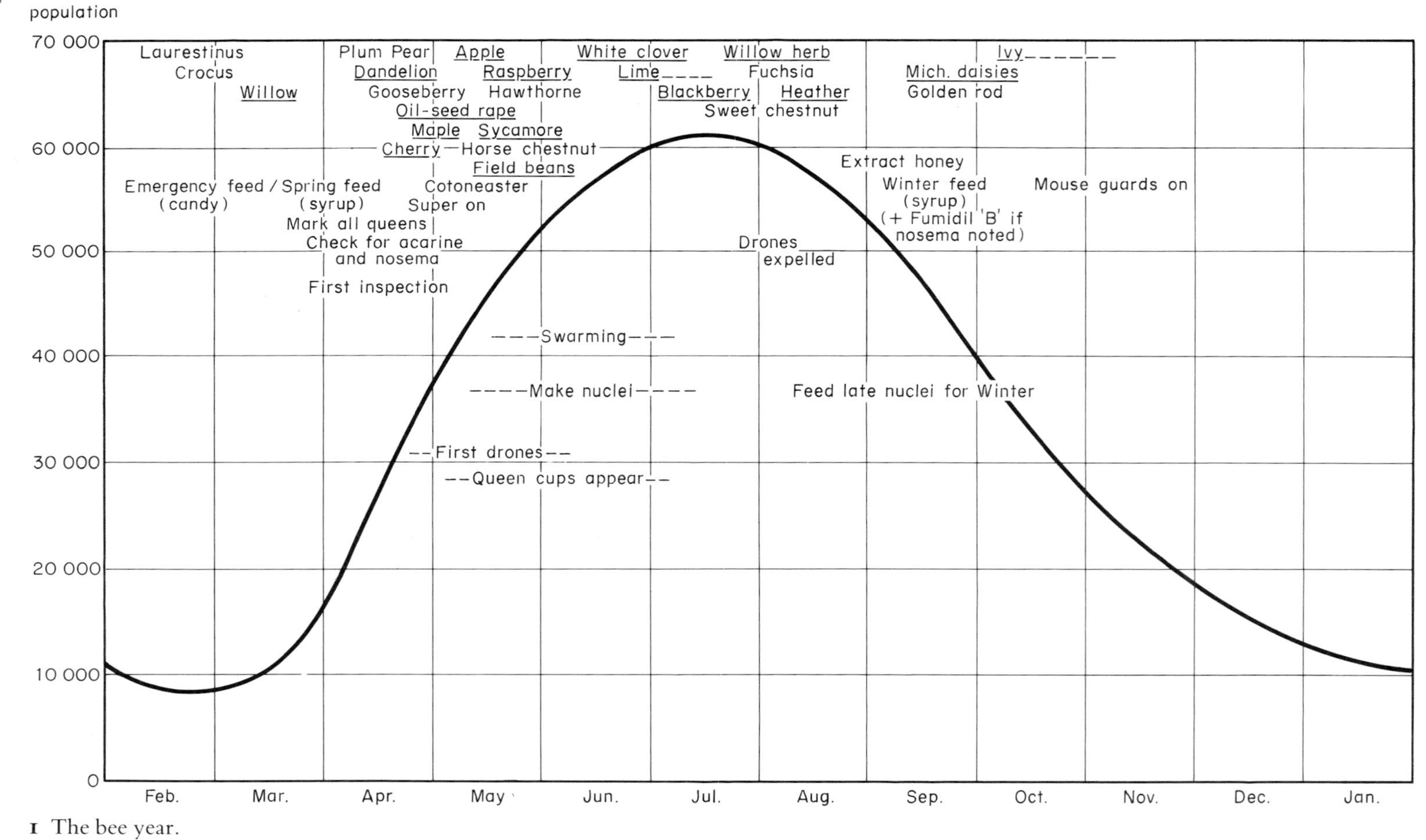

1 The bee year.

this in an out-apiary, perhaps visited only five or six times a year. At home the family cat snoozes on hive roofs, but woe betide the dog if he should desecrate a hive! Bees are very clean creatures and I saw what happened to our black Labrador some years ago – he hasn't done it again since! If the site be well chosen and where necessary screened by a hedge, woven wood fencing panel or even quick-growing tall plants, you can keep your bees anywhere. Lots of people keep them on flat roofs in the middle of London. Are hives very expensive? Yes, but it is possible to make them yourself, or buy them second-hand. Even beekeepers grow old eventually, and there is a great deal of valuable material in attics and garden sheds, or even in the middle of patches of nettles at the bottom of gardens. Why not try a small-ad in your local paper? 'WANTED, hive and beekeeping equipment.' There can be snags, of course, but none that can't be overcome. Making a hive is tricky; for most people by far the best way is to beg, borrow or buy a brood box and copy it, using an empty frame to make sure of a bee-space around it as you work.

2 A fourteenth-century wicker hive, as shown in the Luttrell Psalter. Drawn by Dorothy Hodges and reproduced with acknowledgement to the Trustees of the British Museum.

The basics

Bees need a dry living space of about $1\frac{1}{4}$ cu. ft. (8 gal. or 36 l) for their nursery or brood, as it is called, plus additional room for storing honey, with some ventilation; they don't care whether it is a hollow tree, a straw basket or an expensive new hive. For the beekeeper's convenience it should be a hive of standard design with the bees building their waxen combs in moveable frames, having a $\frac{1}{4}$-in 'bee-space' all round the set of frames, between them and the box. If the space be less than this, they will glue everything down with propolis; if more they will build wild combs of beautiful but eccentric shape and the whole mass will be immovable. Get it just right and combs on wooden frames can be lifted out and manipulated at will. Above the nursery (or brood box) there is a queen-excluder, a metal screen with slots 0.165 in wide so that worker bees can pass freely through to store surplus honey (for you) in boxes called 'supers' placed above. Drones and the queen are too large to pass through, so no eggs get laid in these top boxes and no grubs are reared in them. In a hollow tree the queen may be laying almost anywhere, and honeycomb will be mixed up with patches of brood comb. At the end of the summer, take the honey above the queen-excluder (never from the brood chamber) and feed sugar syrup in September, if necessary, to top up the bees' winter stores.

First steps

This book attempts to describe the first steps a beginner should take, month by month, and

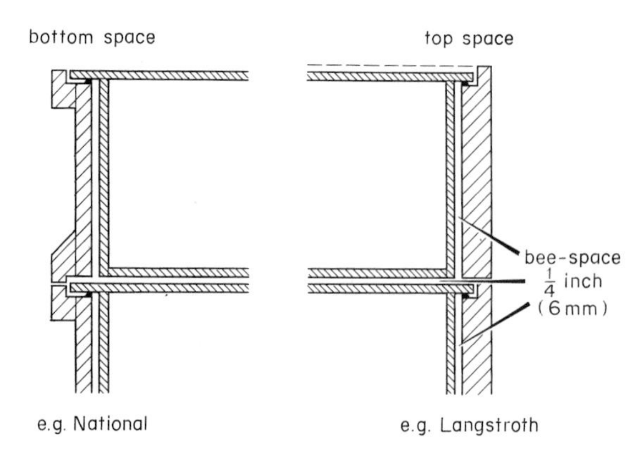

3 The all-important bee-space.

goes on through possible development to commercial or semi-commercial beekeeping. Be content to make haste slowly, and do not take on more than two or three hives until you have gained at least two years' experience. You are bound to make mistakes and it is better to make them on two hives than on twenty.

When and how to start

Without any doubt the answer is late autumn, after joining the nearest bee club (address from Citizens' Advice Bureaux). Most bee clubs have a programme of lectures, film shows and social functions during winter, and also courses of instruction for beginners in January, February and March. They will probably also have a lending library of bee books, but above all they will give you the opportunity to talk about bees with experienced beekeepers. Your county beekeeping association will certainly have its own regular journal or newsletter, but in any case a subscription to a national bee journal will be worth while. (See addresses in Appendix B).

Although the hope of honey in the first season is attractive, many prospective beekeepers have been put off for life by the difficulty of managing a strong colony with no previous experience. On the other hand, a four-frame nucleus with a young queen is very gentle and easy to handle, and as it grows in strength the new beekeeper is growing in confidence, experience, practical knowledge and manual skill. Starting with a swarm is by far the cheapest method, but swarms cannot be guaranteed and in competition with other newcomers one might lose a whole summer. Moreover some swarms will be headed by old queens who may fail during winter and leave an inexperienced owner with a difficult problem. It all depends on the individual, but a very strong recommendation is to begin by buying a four-frame nucleus from an established beekeeper in your district, for delivery in early summer with the promise of a visit a few weeks later for 'after-sales service'.

BEES AND NEIGHBOURS

The main points which put so many people off keeping bees (apart from fear of stings) are the possibility of neighbour trouble and perhaps also the thought that bees are creatures of the country and should be kept on a farm or in a large orchard, at least a mile or two away from the nearest house. On the second point the truth is that today an urban or suburban site is usually better for honey production (on a small scale) than a farm. More people have money to spend on gardens than ever before, and honey gathered from a variety of trees, shrubs and flowers has more flavour and character than most honeys from a single crop. On the other hand, the destruction of hedges to create larger fields, and more efficient farming practices generally, have reduced the honey productivity of agricultural land, except for oilseed rape and possibly field beans, dealt with in Chapter 3. The main purpose of this section is to gather together all possible methods of avoiding neighbour trouble over bees when kept in a village or town community.

If the garden is large enough, then it may be possible to locate the hives so that neighbours will not even be aware that you have bees, and there will be no problem. However, most people have small gardens, or none at all, and in any case it is best for the sake of the beekeeper's family to do everything possible to keep bees good-tempered.

Siting and screening

In general terms it is more important to study the welfare of your family and neighbours than the supposed needs of the bees. Whether a hive entrance faces north, south, east or west is unimportant. Is there a place in your garden where hives could remain almost unnoticed, perhaps between a garden shed and a fairly tall boundary hedge? Can you spare an area perhaps of ten feet by nine feet which could be screened with woven wood panels, a quick-growing hedge or even a thick row of Jerusalem artichokes or stick beans? The latter would at least be most effective at a time in summer when screening is most needed. A quick-growing evergreen climber like a decorative ivy or perpetual honeysuckle on open wire mesh can be effective. Screening should be not less than six feet high to make bees fly up to get away, and deter them from cruising near ground level on approach to landing. Above all, hives should not be placed where the flight path will cross a much-used footpath.

The bees themselves

As already explained, it is always best to start with a four-frame nucleus in early summer rather than a powerful second-hand full hive. The nucleus will give no trouble at all for the first summer, and as the weeks go by the possible fears of people next door will die away, especially if you invite them or their children to 'come and see the bees'. The expense of a spare veil will be well worth

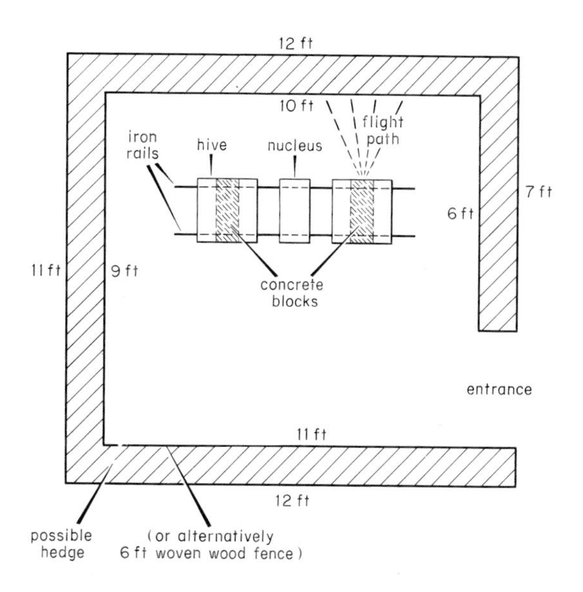

4 A small home apiary.

while. However, a nucleus expands into a full hive by late summer, when neighbours are likely to spend most time in their gardens, and, to be realistic, some bees are better-tempered than others, and you might have a stock of the 'others'. Genetic bad temper is rare, but sometimes no methods of management will deal with the trouble and the only answer is then to requeen (get an experienced beekeeper to help you do this). Some books suggest that the benefit of requeening comes only five or six weeks later, when the new queen's offspring begin flying. My own experience is that a great change can take place within 24 hours, and I am convinced that the quality and quantity of 'queen substance', i.e. pheromones secreted by a queen, play the major part in this.

Environmental factors

There are a few times when even the nicest bees are short-tempered and best left alone. For example, if a promising nectar flow is cut off by a sudden drop in temperature, or a main honey flow (say from an avenue of lime trees)

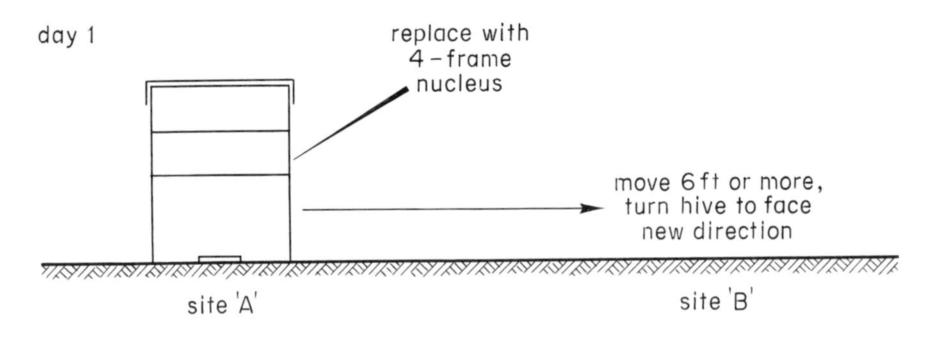

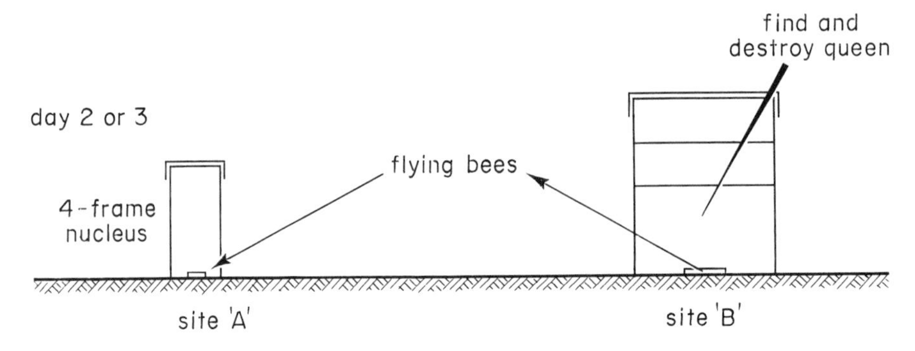

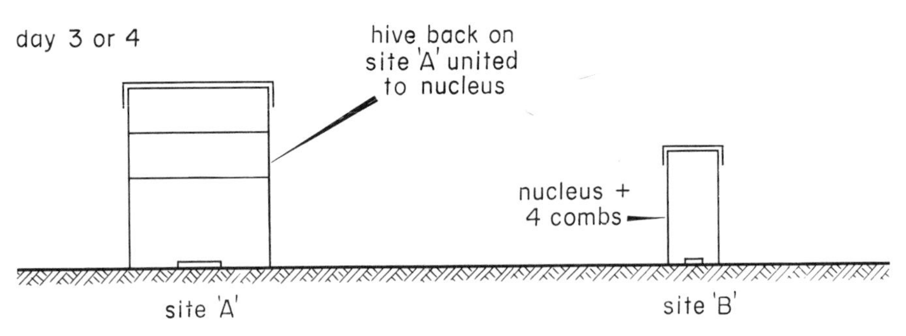

naturally comes to an abrupt end, then thousands of foraging bees are frustrated and readier to be aggressive towards anyone touching or working very close to a hive. If there is a thunderstorm, even in the distance, bees can be bad-tempered and are best left alone.

Sometimes a hive will become bad-tempered as it runs short of food; I have noticed over many years that a strong colony, or even a nucleus, can be irritable when short of food. If a colony develops bad temper, try feeding with warm syrup the evening before you intend to manipulate. A full colony should never have less than 10 lb food reserves at any time. For a really bad-tempered hive, re-queening may be the only solution; see Fig. 5.

Another important factor is whether the

5 Dealing with a vicious colony. *Day 1*: Move vicious stock, replace with nucleus having *young queen*, two frames of brood, two fairly empty frames. *Day 2 or 3*: Older, flying bees now with nuc. Pheromones from young queen usually mollify bad temper. Bees in old stock now fewer and younger; queen easy to find and destroy. *Day 3 or 4*: Bring hive back to site 'A' alongside nuc. Smoke both, open up and take out four frames (two from each end); wide-space remainder to expose to light and air. Take four or five minutes over this, pick up to examine, put down, smoke in between to confuse. Smoke nucleus, take out frames slowly one by one, hold up, then intersperse with frames in main hive. Blow smoke down between frames and close up. If desired, put one frame of emerging brood among the three remaining frames into the nucleus, shake in some young bees and replace on site 'A'. Give queen cell next day.

bees are accustomed to seeing humans near their hive most days. For many years I have observed the calm behaviour of bees in my own garden (urban Torquay), where my wife and I work around the hives all the time. The contrast between this and an out-apiary visited infrequently is most noticeable, and some years ago I wrote in the *British Bee Journal* that my bees recognized me and knew that I was friendly because I talked to them. As expected, this brought facetious comment and letters to the editor, but also confirmation from other beekeepers. The truth is that almost any moving object near a hive will be investigated by bees from a strong stock, and if consistently found to be harmless will be accepted as normal. Thus at an out-apiary one can set up a scarecrow four or five feet in front of a hive, taking care that it will move in a breeze. An old overall or even a long scarf fastened to the top of a six-foot pole will be investigated by bees when seen to move, but lack of response and no harm from it will lead to acceptance, and on your next visit you will notice the difference. It is not necessary that you should resemble the scarecrow! If you do not spend much time in your home garden, then set up this device there too.

Bees flying for water, especially in March and April, can be a nuisance around a neighbour's fishpond, pool or even garden tap. The answer is to provide a source of water for them at home before the end of February, or they may acquire the habit of going next door and stick to it, ignoring your special water supply given too late. A tray of wet peat placed in a sunny corner can be very effective, as well as most useful to the bees.

The smell of cut or crushed vegetation can alarm bees, probably because it suggests a predator working nearby. Cutting the grass in the immediate vicinity of the hives is best done early or late when few bees are flying.

Management points

Dress is important; apart from veil, gloves and bee-suit tucked into wellington boots should be worn to give actual protection and confidence. Animals like bears and skunks with brown fur or woolly coats are ancestral enemies of bees – do not dress to resemble them. Clean, white overalls are best. Bees have a sense of smell and associate sweaty clothes (or even scented hairspray) with animals that might be hostile. Do not, therefore, open up and examine a hive straight from a hard day's work in the garden or from your hairdresser. Hands washed with Lifebuoy soap (giving a weak carbolic smell, itself a bee deterrent) are less likely to be stung than hot, perspiring hands. Even a wristwatch strap (especially if leather) can be an irritant and focus for stinging when working a hive. All this is important to family and neighbours, as the odour of just one or two stings carries the message (via the alarm pheromone) that an enemy is at hand and needs to be repelled. Your neighbour over the fence may suffer from your carelessness.

The timing of hive inspections is obviously important. A commercial beekeeper operating hundreds of hives in several scattered apiaries has no option: the work is there and has to be done, but a beekeeper with two hives in the garden has more choice. Obviously a warm Sunday afternoon in June when the people next door are sunbathing is not the best time to open up a hive. If you have close neighbours don't open up hives when they are working in their garden just the other side of the fence. On the other hand, there are advantages in working bees in good weather between 11 a.m. and 3 p.m., or even on a mild evening, so long as many bees are flying freely. Correct use of smoke is the best way of controlling bees that might otherwise get out of hand, using as fuel wooden shavings, old soft, dry, rotten wood, rolled-up hessian sacking or, as in New Zealand, dry pine needles. Corrugated paper by itself gives too hot a smoke, but rolled up with strips of old sacking is fine.

Avoid spilling honey, syrup or fragments of wet wax. Combs should not be left exposed for more than a few minutes. The unusual presence of such food can easily set up robbing, with thousands of bees flying around hunting for the source, even going in at windows next door. The actual extraction of honey, when done in July or early August, needs very great care if bees are to remain unaware of it, and there is much to be gained by extracting in September or even October.

Psychological and diplomatic points

It is often said that a jar of honey works wonders, and so it does, especially if well-timed. If you hear a child next door coughing at night, offer a jar of honey as a remedy. However, much more than this can be done, and I would suggest you try to enlist the interest and support of immediate neighbours – those who are obviously going to know anyway that you have a hive of bees. Discuss with them the advantage of pollination, the better crops of all kinds of fruit, of beans and other vegetables. If the neighbours have children, get them interested; have at least one spare veil and show them the inside of a hive. Explain that the bees have tiny claws on their feet which can get entangled in hair, and that they do not seek to sting unnecessarily as they die in the process, unlike wasps (black and yellow stripes) which can fly away and sting again. It is seldom indeed that a bee will sting when away from its hive, except by accident.

A difficult hurdle is the way bees will alight on washing and leave a small brown spot, sometimes on the collar of a clean white shirt! This spot consists of pollen husks which have to be evacuated after the contents have been digested by young bees, and facetious attempts

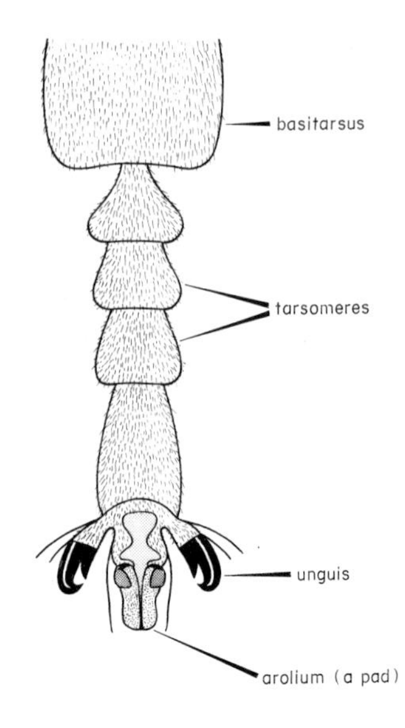

6 A bee's foot, showing the claws that can trap a bee in your hair.

to ascribe the soiling of washing to high-flying aircraft do not always go down well. Fortunately the nuisance is significant only in spring and early summer, when flying range is restricted by cool weather. Newly polished cars near the house can also suffer from pollen husks, which make beekeepers' cars easily identifiable at these times. There is not much that can be done about this, except perhaps to set a limit of two hives at home, so located that neighbours' washing is not in the main flight path.

If a beekeeper wishes to expand to more than a couple of hives, it might be prudent to start up a small out-apiary, not too close to houses, where up to a dozen hives might be kept without neighbour trouble, and where troublesome hives could be moved at short notice if necessary. But even the smallest of gardens could certainly take a couple of nuclei, so that the advantage and pleasure of walking out of the house and watching the bees at work could be retained. A couple of hives can be kept on a flat rooftop even in the centre of London and do quite well, with neighbours often unaware of their existence.

WHAT TO BUY

Personal equipment

This is expensive but very necessary, and new prospective beekeepers should not rush in and buy a full outfit until they are quite sure that they mean to go on despite having received the occasional sting. As a spectator in the first instance at apiary meetings, just a light veil with hands kept in pockets will probably be enough, and even the veil could be borrowed, or home-made. A pair of ordinary gloves will be protection enough for holding a frame of bees handed round to demonstrate stages of brood, and the difference between workers and drones and between sealed honey and sealed brood. However, when opening up a hive it is important to be well protected, and the regulation outfit of a white boiler suit tucked into wellington boots is sensible. The veil must be tucked in under the suit, and should also have a net or gauze panel standing out away from the face; gauntlet gloves are also needed, and then one can face any hive with confidence. The light veil will not be wasted and will always be useful in the garden if weeding close to the hive on one of those days (hopefully rare) when bees show their lack of understanding of gardeners.

A good smoker will last for so many years that it is wise to get the best available, made of copper and of a sensible size. My own was bought 20 years ago and remains a faithful friend, though dented and stained. There are two types of hive tool, as illustrated in Fig. 15, both equally useful, and a slot on the back of the smoker to hold this tool when not in use is well worth making, as is a wooden plug fastened by a wire to the smoker itself, so that the chimney may be blocked at will. Usually the fuel will smoulder on for a good hour after use, if the chimney is blocked, and if you visit more than one apiary can be coaxed into life and smoke at the next call by pulling out the plug and working the bellows. By the same token, it is important not to leave such a smoker in a car boot or garage, but always to leave it outside after use, as a sensible fire precaution.

There are two schools of thought as to whether or not gloves should be worn, but by beginners they most certainly should. Later on, bare hands and sleeves rolled up will impress others, and also impose the discipline of very careful handling, but when opening up a number of hives the fingers get sticky with propolis, making it difficult to pick up a queen should this be necessary. So in the end one comes back to gloves, taking them off for queen-handling, queen-cell implantation and delicate work.

Bee hives

In Britain these come in various sizes and patterns, and the main point here is that they are for the convenience of the beekeeper, not the bees. Bees are just as happy in a hollow tree or straw skep as in an expensive modern hive, but the advantage to the beekeeper of moveable frames in a series of square or oblong boxes is overwhelming. Any modern hive consists of a nursery and a pantry separated by a queen-excluder, a grille through which workers can pass with ease, but not drones or a queen. Within the nursery the queen can range over her brood combs in one (sometimes two) boxes, and above the excluder the honeycombs are housed in a series of boxes called 'supers'. This way the honeycombs are never bred in and remain clean; they can be used over and over again for twenty years or more. In the brood chamber, combs have a limited life, cells getting smaller as they are lined with successive

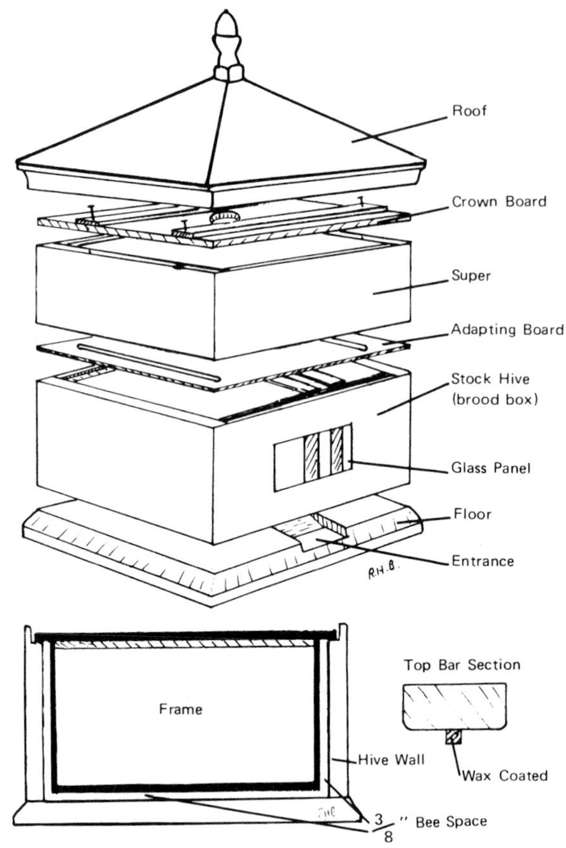

7 The Woodbury hive, 1860 – drawn from illustrations published by Geo. Neighbour & Sons in 1865.

cocoons and pupa cases. In particular, the cell bases thicken as layers of neatly wrapped faecal pellets are deposited by pupating larvae. Baby bees (more accurately larvae) do not have their nappies changed but retain waste matter until pupation, when it is shed in one go. In many countries, combs are not renewed so often as in Britain, and the bees will break down the cells and rebuild them. However, this can lead to higher levels of endemic disease, and there are definite advantages in regular comb renewal, despite the expense. In nature the bees themselves do just this, and more than once in Africa and in Britain I have found a wild colony (in roof space or hollow tree) moving laterally or upwards as space permits, and fairly regularly building new combs and abandoning old ones every year to the extent of about one-third or one-quarter of the nest. The abandoned combs, black with age and use, are eaten by predators.

Over most of the beekeeping world the rectangular Langstroth hive is used, but in Britain most beekeepers (over 90%) use either WBC or National hives (both taking the same British Standard frames). The Scottish Smith hive is simpler in construction and uses similar frames, but with shorter lugs. The British Commercial hive is slowly gaining in favour in the UK and has the advantage of being compatible with National supers, having virtually the same external dimensions; it has

8 Exploded view of National hive.

9 National hive on stand, with double nuc. box.
10 National frames: new plastic frame with wax foundation next to old brood frame fitted with Hoffman converter clips, in front of double nuc. box.

11 WBC hive.

Table 1. Gestation periods (essential to memorize)

	Egg		Open brood		Sealed brood		Total
Queen	3	+	5	+	7	=	15 days
Worker	3	+	6	+	12	=	21 days
Drone	3	+	7	+	14	=	24 days

Table 2. Duties of worker bees of different ages.

Age of worker	Duty
0–4 days	Cell cleaning and incubation
3–12 days	Feeding larvae
About 4th day	Power of stinging fully developed
Between 6th and 10th day	First nursery flights, around midday
6–15 days	Wax-making and comb-building
8–16 days	Reception and storage of nectar; packing of pollen in cell
14–18 days	Entrance guard, debris clearance and funeral bearer duties
19th day	Begins to pay attention to bee dances
18–30/35 days	Foraging for honey and pollen
25–30/35 days	Collecting propolis

larger brood frames which provide 7,000 cells each, as against the BS 5,000, so that five Commercial 16×10 in frames give as much room as seven smaller Standard frames. Thus a full box of 11 British Commercial frames provides the same brood space as 15½ British Standard frames. Modified Dadant frames are even larger, but not so easy for beginners to handle.

Choice of a hive is a personal matter, but it should be borne in mind that in Britain National equipment is easier to buy or sell second-hand, and nuclei are also much more easily obtainable on BS frames, which will only fit into WBC and National hives. WBC hives are perhaps more attractive when standing white-painted in the garden, but Nationals are more easily moved.

Whatever choice is made, be consistent and stick to it; hives in the same apiary with non-interchangeable parts are a great nuisance.

DRONES, WORKERS AND QUEENS

A normal, healthy colony of bees will have a population varying from 10,000 or less in winter to 50,000 or more in high summer. There will be a queen and workers only for most of the year, but for the summer months also drones, usually just a few hundred. The queen can live for three years, sometimes even four, but is normally fully efficient only in her first two years. The lifespan of a worker, like that of a car, depends more on mileage than age; at the height of summer with long working days it may be as short as five weeks, but in autumn, winter and early spring up to six months. Drones normally live only three or four weeks and are usually thrown out at the end of summer; earlier in a bad season or if a newly mated young queen is present. Recognition of adult queen, drone and worker bees, and the cells that they are raised in, is easy, but the recognition of brood at various stages is harder, and it is necessary to learn and remember the basic facts of bee life if one is to 'read a comb' and understand what is happening when 'going through a hive'. (See Table 1.)

Worker

The jobs a worker bee performs in its short summer life are many and varied. Table 2, showing the duties usually carried out by bees of different ages, is only an indication. If need should arise the older bees can regress and carry out the tasks of younger bees, which in turn can take over the duties of older bees earlier than usual should this be necessary. In late autumn and winter worker bees are mobile food reserves and carry both protein and fats stored around their glands and in layers of so-called 'fat bodies'. They consume stored carbohydrates to maintain warmth and bodily activity, and use the other food reserves in spring to manufacture brood food when the queen resumes egg-laying on a larger scale.

There is obviously much overlapping, and many bees may omit certain jobs altogether. The working life of a foraging bee is lengthened if confined to hive by bad weather for several days.

Drones

Drones, the male bees, arise from unfertilized eggs, which pass the spermathecal duct without receiving sperm. Thus drones have a mother but no father. This is called parthenogenesis.

A queen will normally lay in drone cells some time in April, seeking out existing drone cells in order to do so, and often bypassing available empty worker cells in the process. If there is no drone comb available, the workers will build clumps of it at the base of existing combs, or tear down areas of worker comb and rebuild it to create drone cells; sometimes they will squeeze drone cells in odd places between the woodwork of frames. This can make the inspection of hives in May a messy business, and also result in areas of good worker comb being spoiled so far as the rest of the year is concerned, with knobbly areas of once-used drone comb thickened with propolis and wax. Figs. 44 and 45, showing composite worker/drone comb to be inserted about one frame in from the side, shows how this may be avoided by giving the bees what they need at the time they need it. Another method is to give space under the brood frame by inserting an eke about 2 in deep, or by using brood chambers deeper than usual. The existing worker comb is then preserved as such, and the bees build extensions of drone comb beneath some of the frames. This may be removed at the end of each season for wax production, or left on for next year. Some of the extension comb built below will be of worker cells, if the colony needs more room.

The production of drones does not mean that a colony is necessarily going to swarm; it seems to be an insurance policy taken out in case the need to swarm should arise later. Young drones normally fly about the sixth or seventh day after emergence, but are not usually virile until the eleventh to fourteenth day. Flying drones are readily accepted and fed in hives other than their own and may join any swarm. Towards August the drones are driven out of hives, even when food is plentiful, but are retained by queenless stocks or stocks with a failing queen. Thus the presence of drones after this time can be a signal warning of queen failure, either actual or expected. Wedmore in his *Manual of Beekeeping* quotes a figure of about 1¾ lb of food needed to raise 1,000 drones, which will in turn consume about 4 oz of honey a day, and suggests management to reduce their numbers. The writer questions whether 1,000 drones, weighing under 7 oz (about 200 g) do in fact eat more than half their body-weight of honey a day in the limited time that they fly, and suggests that 1 oz a day (28 g) would be nearer the mark. In any case, seen against the hive's total annual budget of about 400 lb honey and 50 lb or more pollen, the expenditure on drones is small and worth while. In this respect perhaps the bees know best. Most experienced beekeepers agree that colonies with plenty of drones seem to produce as much honey as those with only a few. (See also section on drones in high summer, page 87.)

Queens

A queen is virtually an egg-laying machine, and it seems doubtful if she exercises any conscious control herself. The evidence is that she is manipulated by the workers who control her feeding. Yet her influence is powerful in many respects, and the temper of an aggressive colony has been seen to change within 24 hours of the introduction of a new queen: presumably this is because of queen-substance pheromones more generously secreted. As a mated queen carries all the genes responsible for worker characteristics, it is she who determines the basic work and behaviour pattern of the colony, as modified by her own pheromone secretions. The actual factor determining whether a fertile egg shall produce a worker or a queen is the larval feeding, and it has long been known that larvae up to two days old taken from or in worker cells and fed by the bees on a richer diet of brood food (royal jelly), will develop into queens. Experienced queen breeders say that the best queens arise from larvae given royal treatment from the age of between 12 and 18 hours, although the difference is not great up to an age of about 30 hours. It has long been debated as to whether royal jelly contains substances additional to normal brood food, or whether it is the quantity given which triggers off hormonal reactions to produce a queen emerging on the fifteenth or sixteenth day instead of a worker on the twenty-first. Recent research suggests that both worker larvae and royal larvae are fed on glandular food throughout, and that the older explanation that worker larvae are 'weaned' at the age of two days and given a diet of pollen and honey thereafter is not valid. The presence of some pollen grains is now thought to be accidental, the determining factor being the ratio of honey to glandular food. On this basis it seems (somewhat surprisingly) that a higher proportion of natural sugars in the diet of one- or two-day-old larvae triggers off activity in glands (*corpora allata*) producing hormones which activate the development of ovaries in potential queens, and also suppress such worker characteristics as barbed sting, pollen baskets, brood-food glands, wax glands and so on. However, I suspect that we do not yet know the full story.

As shown in Table 1, a queen normally emerges on the fifteenth day after the egg was laid. Some very experienced queen-rearers who incubate queen cells artificially have found that a slightly lower temperature (say 90°F instead of 92–95°F), will slow down development a little and produce marginally better queens, emerging late on the sixteenth day. Possibly this is why natural swarm cells are constructed on the outside of the brood nest.

Young queens fly more frequently than is generally supposed during their first few days, before mating, and seem to do so just after the one o'clock news. On so many occasions I have paused to watch the activity of mating nucs on the garden wall at home, after hearing the news, and seen young queens only two days old walk out and take flight, with workers ignoring them completely. These queens usually return within four to seven minutes, presumably after exercising flight muscles and learning the visual pattern for navigation homeward. When they are about a week old, their workers show much more interest, and when a queen is out, some workers will be found fanning, with Nassanov glands exposed, setting up a navigational beacon of pheromones to guide a mated queen homeward-bound. (Pheromones are external hormones, which convey important messages by their smell, from worker to worker, from queen to worker or from queen to drones.) Mating usually takes place when a queen is seven to twelve days old, so long as the air temperature is at least up to the low sixties Fahrenheit (over 16°C). Queens mate with up to seven or eight drones, usually on one or two flights on the same afternoon, but sometimes on successive days, and return to their hives with the 'mating sign' derived from the genitalia of their last partner. A mated queen is treated with much more respect than a virgin and fed 'on demand' thereafter, so that egg production may begin within two or three days. Such feeding also stimulates secretion of the queen-substance pheromones which have such a profound effect on the mood and work pattern of the colony.

After mating, the queen does not fly again, except to lead out a prime swarm the following year, or more often the year after. In preparation for swarming the workers will feed her less, to slim her down for flight by greatly reducing egg production in the tubes of her ovaries. Presumably this reduction in food also affects her pheromone production, reducing the inhibition on queen-cell production and reinforcing swarming preparations. A significant drop in egg-laying is one of the points noted by sharp-eyed beekeepers looking for signs of swarming intentions. Thus the decision to swarm is finely balanced, and can be reversed by the onset of a bountiful honey flow and more generous feeding of the queen. The fact that more bees are flying in a honey flow also means that the hive is less crowded, and that the circulation of queen substance from bee to bee is less impeded. Thus, not all colonies with swarm cells necessarily swarm, and the cells may be torn down by the bees themselves. Obviously also, an older queen may be producing less queen substance anyway, so that sometimes a colony may swarm even when it has not yet filled even the brood box. A satisfactorily mated queen may have something like seven million spermatozoa stored in her spermatheca, making her capable of laying up to one and a half million fertile eggs over the next two or three years.

Many books refer to the 'piping' or very high-pitched sound uttered by virgins in a hive. I have made tape recordings of two quite separate and distinct sounds, viz.:

a *Quarking* – a low-pitched, resonant sound, 'Quark, quark', repeated up to seven or eight times, made by virgins still in their cells. One such recording was made from a ripe queen cell taken from a hive and laid on a microphone, about five minutes before she had bitten out a disc at the cell end in order to emerge.

b *Piping* – a very high-pitched squeak uttered by virgins within the hive when they have emerged and are on the combs. This sound makes workers 'freeze' and remain motionless on the combs for several seconds, presumably helping the virgin to reach and attack other queen cells. A most interesting experiment at Rothamsted research station about 1972 demonstrated this effect on workers by artificially producing this sound within a large observation hive. The sound of piping seems to stimulate virgins still in their cells to reveal their presence by 'quarking', leading to their own destruction by the piping sister.

Two

Spring – the great awakening

SPRING MANAGEMENT

One may well ask, 'When is spring?' There are several possible answers, but to the beekeeper it surely is the time when queens are stepping up their egg production and the number of young bees emerging increases every day, so that the population is no longer falling and will soon begin to rise again in readiness for summer still many weeks away. The exact date when this happens may vary from place to place or even from hive to hive, but mid-February in the south of England and two or three weeks later in north-east England and Scotland would probably be an average date.

In some years there may be the occasional hive which is short of food as early as February, perhaps because of unnoticed robbing by wasps or other bees in late autumn, or a very mild winter may have encouraged more breeding than normal with unusually high food consumption. Good autumn management with attention to: *a* adequate feeding; *b* restriction of entrance against robbing; *c* mouseguards; and *d* adequate top ventilation should have prevented this, but perhaps for some good reason these precautions were not taken, the hive feels very light when hefted, and the owner is worried. What can be done about this in February? Perhaps the first point is that one needs to be more precise than just feeling how heavy a hive is; hives may vary a good deal in their empty weight and unless you have recorded it, you are unlikely to remember. The best plan is to use a spring-balance measuring up to 55 lb (25 kg) and methodically weigh each hive a week or two after autumn feeding has been completed, when the syrup has been evaporated and stored. This is quite easily done by hooking the balance under one side of the hive and pulling until it comes up just an inch or two off the stand. Then repeat this on the opposite side and add together the recorded weights to get the full weight. (See Fig. 82.) Absolute accuracy is not important; a figure within a couple of pounds is good enough. If the weight at the beginning of February is only 15–18 lb down on the weight at the end of September, then all is well. If the loss in weight is 25 lb or more, then a slab of candy weighing 5 lb ($2\frac{1}{2}$ kg) would be a wise precaution, to help the colony through the next month until a visual check on the combs may be made on a mild day, and a gallon of thick syrup safely given.

The first inspection

We usually regard the period from October to

March as the 'off season' so far as opening up hives, lifting out frames and manipulations generally are concerned, but a limited inspection on a mild day in March can be justified on the grounds that we can only help our bees if we know what help they need. Just two questions have to be answered at this time of year: *a* have they a laying queen? and *b* have they enough food? In some cases the answers may be obvious; for example, if the bees are flying freely around midday and taking in massive loads of pollen, then all is well with the queen. If the hive still feels really heavy when hefted, then they have enough food. Possibly about one hive in three will either feel light, or show little flying activity with not much pollen going in, and in these cases some action is required. As I have mentioned, the 'feel' of a hive may not be accurate enough, and the method of weighing with a spring-balance is strongly recommended.

At this time of year a minimum of smoke is necessary, just a couple of gentle puffs in the entrance to drive in any guard bees, then a two-minute wait before continuing. If the hive has been wintered on a brood chamber, queen-excluder and super, first take off the super plus crown board and place it diagonally over the inverted roof. If there are bees in this super they will take no harm, and unless the excluder is faulty there will be no eggs or brood there to chill. Assuming that you have opened up a hive before, the only difference now is that a warm quilt should be laid over the frames as the queen excluder is removed, and only the outermost frame on one side actually taken out, no others even being lifted. At this time of year there will be no bees on the outer frame, so gently lift it out and lean it against the outside of the hive. Now use a hive tool, and with a gentle, levering movement separate the next frame from the third, pulling it over to the end and noting how much sealed food there is on it. Repeat this with the next frame, and possibly one or two more, until the brood nest is reached. If the bees are clustering tightly it may be necessary to drive them down with a puff of smoke in order to see the brood. You do not need to see the queen, just a patch of eggs or brood to convince you that she is present and laying. Finally, push back the frames, replacing the outer one, and close up. There should be at least one outer frame heavy with sealed food, as well as a layer of food perhaps two inches deep at the top and sides of the central combs. Look carefully for any sign of soiled combs, where bees may have defecated inside the hive; also look for stains on the front of the hive itself.

Food shortage

It has to be remembered that food is used up very fast from about the second week in March, when a good queen gets into lay again on a large scale; much more food is needed to build up the body tissue of thousands of developing larvae than to keep the overwintered bees alive. Don't just say, 'They have come through the winter OK,' when you see a few bees flying in the first week of March. The period from mid-March to mid-April can be critical, and if in doubt the answer is to feed a gallon of thick syrup. If it is earlier than mid-March, or if the weather is hostile, a large slab of candy over the feed hole might be better, as previously mentioned, since syrup stimulates bees to fly but candy does not. A specific shortage of pollen would be unusual in the English spring, but may occur in some areas. This can be dealt with by making up an 'artificial pollen' patty. There are many recipes for this, but most contain soya flour, sometimes with skim-milk powder and dried yeast. For most practical purposes, the addition of one part of natural pollen to three parts of soya flour, made into a moist patty with honey and placed directly over the brood nest on the top bars of the frames, will provide a protein food which the bees will readily use if there is a shortage of natural pollen.

Queen failure

If, in the second half of March, there is no evidence of eggs or brood at all, there is the possibility that the queen has died or gone completely off lay. If there is evidence of drone brood (high-domed cell cappings on worker cells), then she has run out of sperm. It will not be possible to do anything until early April, but make a note to insert a frame of eggs and young larvae from another hive at the next inspection. This will enable the colony to raise a queen, with a 50–50 chance of mating, and in any case will balance up the population before you take further action.

Disease

The so-called 'spring dwindling' is not itself a disease but a symptom. Assuming that food stocks are adequate, nine times out of ten the cause is either queen failure or nosema disease. Bees are more likely to soil combs after long winter confinement, especially in early spring when activity inside the hive increases but cold weather prevents free flight. A few over-wintered bees infected with nosema will foul the combs with spore-laden excrement, and healthy young bees are infected as they attempt to clean up, or as the cluster spreads and soiled wax cappings are chewed off to get at the stored food. This causes a rapid spread of infection, and as infected bees have shortened lives, with under-developed brood-food glands, colony growth is badly checked. Fortunately there is a specific remedy, Fumidil B, either provided by local Beekeeping Associations or on sale from bee appliance dealers. Fumidil B is an antibiotic which effectively controls nosema disease, and is best given in winter syrup fed in September if trouble is suspected. However, it is also effective as a spring medicament, so now proceed as follows:

a open up the infected hive and remove side combs not covered with bees, shaking off any bees as necessary;
b put in a dummy frame on either side of the cluster to hem them in and retain warmth;
c put on a contact feeder with a gallon of syrup plus Fumidil B.

This deals with the infected bees; now the soiled combs have to be sterilized with 80% acetic acid. If you have more than one hive infected, the usual method is to make a stack of boxes of combs with a saucer containing a third of a cup of acetic acid and a blob of glass wool between each box. The whole should be sealed top and bottom to prevent fumes escaping, with wide adhesive tape such as Sellotape, or masking tape, around the junction of one box with another. (See Fig. 12.) If there are any brood combs from colonies which have died out, or any suspect combs from store, it is convenient to treat all together at the same time. Let the fumigation proceed for about a week, then air the combs thoroughly for a day or so (in a greenhouse, for example, or where flying bees will have no access to rob), and replace the treated food combs so that the infected colony can expand with no risk of re-infection. The dummies should then be removed, or placed outside all the combs.

Acetic acid must be handled with care, as it is corrosive and attacks metals. Metal ends should be removed from frames and replaced afterwards. It does no harm to honey or pollen, but effectively kills nosema spores, and any other unwanted pests like wax moth larvae at the same time.

Nosema weakens colonies more often than it kills, and sometimes infected stocks will appear to recover by themselves as the weather im-

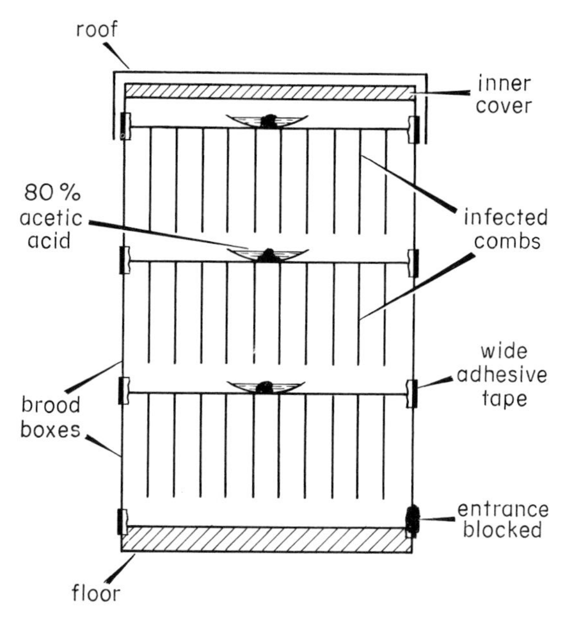

12 Sterilizing brood combs.

proves and bees fly freely, able to defecate outside the hive. However, such colonies usually build up too late to gather a honey crop, and perpetuate the infection unless treated with Fumidil B in their winter food. The foregoing information, condensed though it is, may put off some inexperienced beekeepers, but many years of experience have convinced me that nosema is the biggest single reason why some colonies survive year after year while producing very little honey. If there is no evidence of nosema, then failure to build up is probably the fault of the queen. If young, she may have been imperfectly mated, or perhaps she is old and shortly due to fail completely; in either case make a note to check more closely early in April and be ready to re-queen.

Warmth

Although ventilation is important in midwinter, there comes a time (usually in March) when the expanding colony with thousands of growing larvae needs more water than it produces as a by-product of food consumption. At this point, through ventilation becomes less important, and heat conservation more important as the brood nest expands, and there are perhaps scarcely enough adult bees to cover the brood nest and keep it warm. Hence the old slogan, 'Keep them warm in March', and the wisdom of putting on a blanket of insulation fibre at this time.

The turn of the tide

When dandelions come to their peak towards the end of April, and the occasional warm day allows the bees to forage freely, the big build-up really gets under way, or should do. As a guide, any normal, healthy colony should have brood on six to eight National frames by the end of April unless there is some good reason why not – for example, perhaps a colony had been transferred from a nucleus box in August and never had time to build up beyond five or six frames by October, or perhaps the colony was a late swarm the previous summer. Having allowed for such factors as these, it is important not to make excuses and hope that the queen may be a 'late developer'. Nine times out of ten it pays to re-queen if in doubt, and to do so early enough to permit the new queen to build up the stock *before* the honey flow, and not *on* the honey flow. This is the time to watch progress closely, and should there be a prolonged spell of cold, wet weather, think of the food needs of that active colony unable to fly, yet with thousands of babies to feed. Be ready to provide syrup if necessary.

Some beekeepers say that by the time the horse chestnut is in blossom the danger of starvation is over, but our climate is so variable that twice in the last ten years this has not been true, and it has been necessary to feed healthy stocks even in May to keep them strong. We cannot alter this climate of ours; all we can do is to be ready to help our bees when they need it.

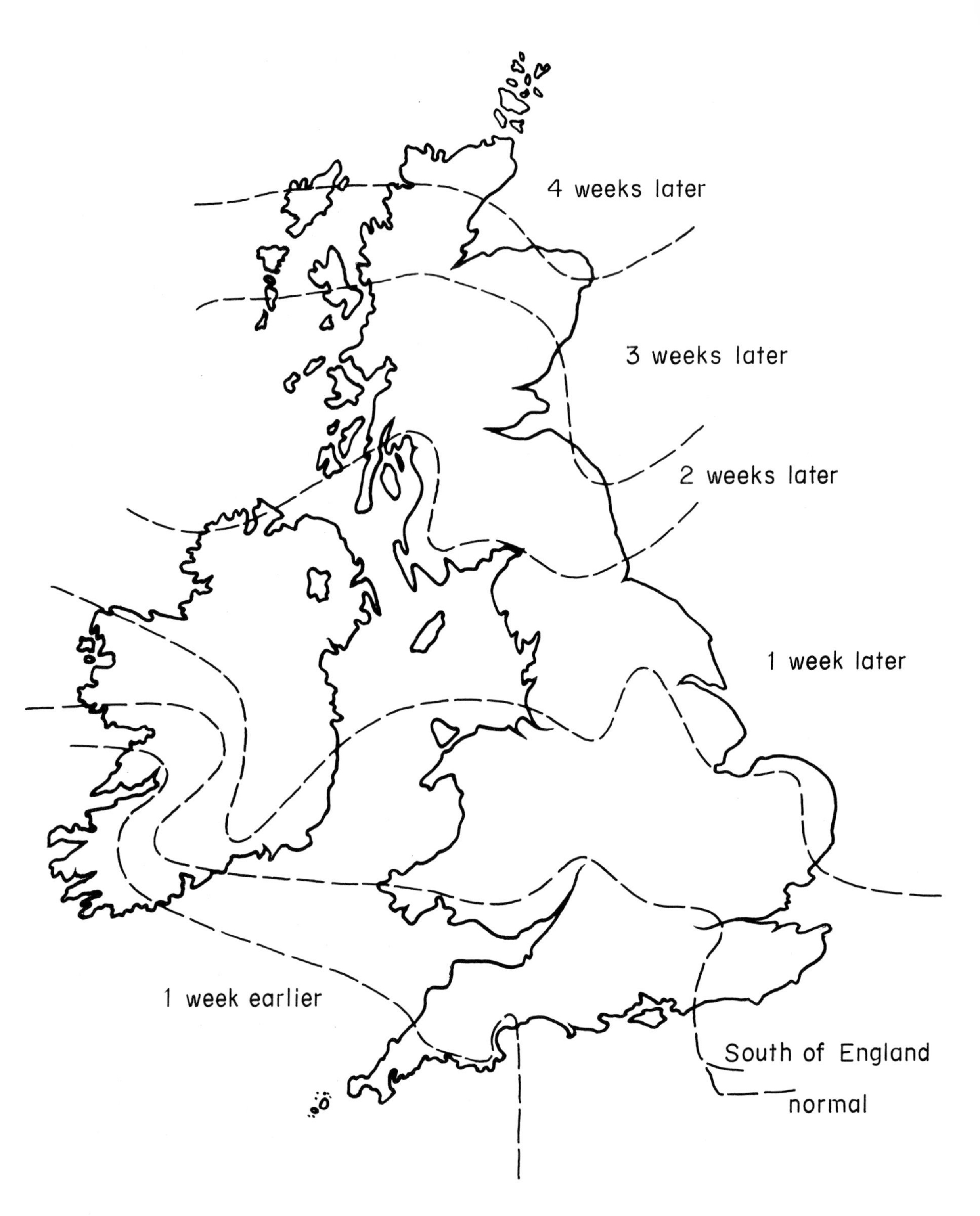

13 April: the arrival of spring. As the contours show, there is a difference of over a month between Cornwall and the north-east of Scotland in April; the difference is a good deal less later in the summer. The dates of flowering of nectar-yielding plants will vary according to the zones indicated. Adapted from *The Gardening Year*, Reader's Digest Association Ltd.

BEGINNERS – MARCH (SECOND YEAR)

At this time of year large numbers of bees will be collecting water to dilute the stored honey and make it suitable for feeding to the larvae. Two years out of three this presents no problem, and when it rains every other day the bees have not far to go for water, but if we have a dry spell between mid-February and the end of April, the problem can become acute and bees may have to fly some distance with temperatures in the low forties (°F), and then return with a load of cold water. They usually do this during sunny periods around midday, but it is touch and go and many bees never get back. Since they have to carry water in their honeysacs, they have no honey or fuel for the return flight and have to depend on the sugar content of their blood. It is rather like coming back from the supermarket round the corner with only the petrol in the carburettor to get the car home. Once they land in a shady spot their body temperature drops rapidly and they are then unable to fly.

You can help by providing a safe water source in a warm corner of your garden, in the sun and out of the wind, a few yards from the hives as bees seem to prefer their water supply to be at a distance, but not more than 20 yards. One of the simplest ways is to use whatever shallow vessel comes first to hand (old stone kitchen sink, inverted dustbin lid, old chicken or pig trough, plastic or metal seed tray) and fill it with water, adding a handful of hay, pieces of bark, a spadeful of moss peat or anything which will give a foothold and help prevent bees from drowning. This is best done in February, otherwise there is a possibility that some enterprising bees will already have discovered the tap outside the kitchen door, and will loyally continue to visit it, ignoring the official bee-watering arrangements, and thereby possibly causing domestic friction. Some years ago I experimented with a wide, shallow trough of water, with some stones at one end, wood in the middle and old, blackened sacking and peat at the other end. Even in weak sun the moist, darker material soon became several degrees warmer than the water by the stones, and 95% of the bees (dozens at a time) showed their appreciation by going to the warm end.

Not all wisdom can be reduced to a short slogan, but 'Keep them warm in March' is still good advice. For most of the winter, damp, not cold, is the enemy, but in March the picture is rather different; on balance there is now more water coming into the hive (carried by bees) than going out, and condensation is not so much to be feared. In fact, if you have a glass quilt you will notice bees licking up the drops of condensation probably to be seen around the edges. In March the brood nest is growing fast and before the end of the month there may be eggs and brood on four to six combs. The bees will be hard pressed to cover these and maintain a temperature of above 90°F over such a large area. They can be helped by a warm but porous covering placed over the crown board. A square of carpet or underfelt will serve, but better still is glass wool cut from a roll of Superwrap, sold in rolls conveniently 18 in wide and 3 in thick for insulating the roofs of houses. If you have to put a feeder on, make a suitable hole in the carpet or glass wool to accommodate it, and ensure minimum ventilation by at least three matchsticks placed under the feeder. Sheets of white polystyrene, often used as packing these days, will also make effective heat insulators. If you have wintered on a wide entrance, now is the time to put in the entrance block, still with the mouseguard over, to give a bee-way about 3 in × $\frac{3}{8}$ in or thereabouts; this will reduce draughts at a time when less ventilation is acceptable.

Another job which can usefully be done in March is to clean or change the floor; no smoke is needed, just a steady pair of hands to lift the hive off the floorboard and set it gently down on the stand alongside, usually replacing it within a minute. If you find the floor to be dry with just a few powdery wax fragments and maybe a dozen or so dead bees, then you have probably wintered successfully. If the floor is wet and foul, then most likely you had insufficient ventilation during winter. If you failed to fit a mouseguard there is about one chance in ten that you have an additional tenant, and the evidence for this will be some dead leaves and grass, slightly larger pieces of wax from combs (some as large as your small fingernail), and mouse droppings. More of this later.

Perhaps most important of all is to form some opinion of the amount of food left in the hive, without opening up yet to have a look. The weight of the hive is a good indication, but only if you have experience. However, by now you have probably made or acquired another hive ready for the summer, so use this for comparison, and if the stocked hive does not feel a good deal heavier, you may have to feed with sugar syrup. Do this in the evening, using thick syrup (2 lb sugar to 1 pt water) with a contact feeder, i.e. a plastic bucket or inverted tin with small holes over an area of a square inch in the centre of the tight-fitting lid. The point of a contact feeder is that in a spell of cold weather the bees may not be able to climb up and over to get syrup from the usual feeder, but can suck syrup from small holes presented at the feed hole, perhaps without even leaving the cluster. (See Fig. 70.)

If there are no bees in the super, and no sealed combs visible, then the empty super could be removed in mid-March in order to get the food nearer to the cluster. If there are bees on two or more frames, the super should be left in.

During March

a Set up a water supply about 15–20 ft away;
b check that entrance blocks are in and cover crown boards with glass wool, blanket, or a carpet square;
c change, or clean, the floorboard;
d estimate food stocks, feed syrup if necessary.

BEGINNERS – APRIL

A word to existing beekeepers. This is the time of year when new members join, and with spring in the air lots of people feel the urge to keep bees. How best can we help them and what advice ought we to give? First and foremost, get them to come to branch meetings and make them feel welcome. Remember that last evening meeting when you noticed a new face but were too busy to go over and speak? It is so easy to get tied up with arrangements and branch affairs, talking with old friends, swapping stories about bees and grumbling about the price of sugar. All branches have committees; why not choose someone who always comes to meetings and make him or her responsible for new members, as an 'extension officer', to borrow a Rotary term?

Now a word to beginners. Take every opportunity to ask beekeepers how they keep their bees; most people love to talk about their hobby and give advice. The great thing about advice is that people like to give it; it costs nothing and you don't have to take it! At apiary meetings explain that you have never seen eggs, larvae and brood and ask to be shown them. You won't need to ask to see the queen; if the demonstrator finds her you will hear all about it. Old hands make mistakes too, and learning from mistakes is the finest education there is. Learning from other people's mistakes is best of all!

Now for starting with bees yourself. As previously stated, the best way by far is to buy a four-frame nucleus with a young queen for delivery at the end of May or first week in June. Do try to have your first hive where you can see it often, in your own garden if possible.

The point of starting with a nucleus is not only that you can more easily learn from it, but also that it is unlikely to be a nuisance in a small garden. By the time it has grown into a strong colony, the screen of runner beans or artichokes will have grown too.

Work for April

a Join your local branch of the County Beekeepers' Association;
b read monthly journals regularly and borrow a book or two on bees from the library;
c order a four-frame nucleus for the end of May/early June (your local branch may be able to advise, or even supply);
d buy a veil and gloves and attend apiary meetings;
e choose a secluded corner of your garden, allow space for two hives side by side as well as for working, and trench around ready for artichokes and runner beans.

BEGINNERS – APRIL (SECOND YEAR)

April is the dangerous month, when stocks can so easily die of starvation. Every year we hear of beekeepers saying in March how well their bees have come through the winter, only to find that they have died out by mid-April, and on opening up there is found a large patch of dead bees with heads stuck in empty cells and tongues still out asking for food. If in any doubt about food stocks, feed a gallon of thick syrup. Bees at their last gasp, needing first aid rather than just feeding, should have warm syrup squirted directly into the empty combs with a plastic washing-up-liquid bottle; I have used this at demonstrations for many years, also with mini-nucs sometimes in summer. If you do not possess a feeder then put a block of soft candy or baker's fondant over the feed hole, or even, in a serious emergency, a 1-kilo paper pack of white granulated sugar. With the latter, one can cut a small round hole the size of a ten pence coin out of one side of the packet, pour in half a pint of water and invert it over the feed hole. The water will stop the sugar from trickling down non-stop into the hive. (See Fig. 83.)

It would be rash to advise beginners to open up a hive in April unless the temperature is up to 60°F, but if you have an experienced friend to help you, a limited operation, with the air temperature over 50°F and no wind, will do no harm. Have an inspection cloth ready, use a minimum of smoke and work a hive tool gently under the crown board at a corner, twisting slowly to ease it up from propolis possibly sealing it to the hive body for the winter; move to an adjacent corner and repeat, trying to avoid any sudden jerks. Blow a soft puff of smoke into the gap and then check that no frames are going to lift. Sometimes bees will build burr comb over the lugs, losing the bee-space, and then stick the lugs to the crown board. The danger is that you may raise the board too quickly and a couple of frames will be lifted up two or three inches then drop back with a crash, and out will come the bees to see who is attacking them. When the board is up, replace immediately with the cloth and roll it back an inch or two. Now proceed as described on page 24 ('The first inspection'). The whole job need take only two minutes and will have done no harm. In any case you can only help your bees if you know their exact condition, and in April you must confirm the presence of food and a laying queen. Bees are almost always good-tempered and slow to take offence in April, and an inspection like this will do wonders for your own self-confidence. If you feel quite unable to undertake this inspection, take comfort from the knowledge that nine times of out ten, bees working with urgency and taking in masses of pollen will correctly indicate a laying queen, and to an experienced hand the weight of the hive will disclose the need for feeding or not.

This is also a good time to rake under the hive and clear away dead leaves and grass, weeding around the stand on a cool part of the day when the bees are not flying. A good plan is to save cinders from a coal fire and spread these under and around the hive and stand. Well-trodden-down cinders are the poor man's tarmac and will prevent vegetation from growing too close. Old engine oil will also serve.

During April there can be a substantial flow of nectar from dandelions and soft fruit bushes, but often winter stores are still being used up. If the super was taken off in March when feeding, replace it by the end of April.

Make a point of looking at your hive as often as possible, especially around midday, and you will notice the steady build-up of activity. Watch the pollen loads going in – you should see large loads of very bright orange from the dandelions, a mid-grey from gooseberries from about the middle of the month, possibly some blue from scilla if there are gardens nearby. From now on there will be nursery flights of young bees in the lunch hour, strengthening their wings and learning to recognize the hive and its surroundings. Remember that eggs laid in April will become flying bees in the second half of May and early June, when there is real work to be done.

In April

a Check that you have a laying queen and plenty of food in the hive, opening up if necessary;

b rake under the hive and spread ashes or gravel around;

c watch pollen going in, bees taking water from your trough, young bees on nursery flights and relate to air temperature. Keep beekeeping notes or diary.

BEGINNERS – MAY

Siting a hive

In all matters to do with beekeeping there are two considerations: what is best for the bees and what is best for the beekeeper. Usually the solution has to be a compromise between the two, and choosing a site is no exception. For the sake of the beekeeper, hives should be put where they will not be noticed too easily; any old hand will confirm that other people seldom get stung by bees when they cannot see the hive alleged to cause the trouble. If passers-by or callers at the door see a hive, then your bees may collect the blame for anything from a wasp sting to a mosquito bite. Whether the hive faces north, south, east or west is not really important, and Brother Adam at Buckfast Abbey has hives in groups of four facing in all directions. If you have a choice, arrange matters so that the sun shines mostly on one of the thick sides of the hive, i.e. a side that carries the ends of the frames. This will help to distribute solar warmth more evenly between the combs. In Britain there is no need to avoid direct sunshine; in fact, hives in a sunny, sheltered spot have an advantage. In the tropics, the shade of a tree can help, especially with Adansonii bees.

For the sake of the bees, avoid a draughty corner where gusts blow into or across the entrance; when foragers are flying in with a heavy load of pollen or water in early spring, at an air temperature of perhaps only 45°F, their

14 Buckfast Abbey home apiary. Note hive entrances face north, south, east and west.

motors may only be running on about two-thirds of their full power, and landing in a cross-wind is difficult at any time. If bees force-land in the grass with no sun to warm them they cannot develop enough power for lift-off and you will find them on the ground, pointing towards the warmth of the hive they just failed to reach. If you have time, give them the 'kiss of life' by warming two or three at a time in your cupped and almost closed hands and breathing slowly on them. Usually within a minute they will start walking and almost at once be able to fly those last few feet home.

A hedge or open-woven wood fence is better shelter from wind than a solid wall, as there are more gusty eddies in the lee of a solid wall. To avoid dripping water and falling debris it is best in Britain not to have a hive under a tree, and for obvious reasons a hive should not face on to a garden path.

Various books make different suggestions as to hive supports, but in my experience the best way is to arrange parallel metal runners on two concrete blocks, so that there is a clearance of at least six inches under the hive floor, to enable a rake or hoe to be used, as well as to allow air to circulate. Never keep a hive resting on the bare earth, or the floor will be damp, and despite treatment may rot after a few years. A pile of bricks is not ideal, and may provide an easy way for ants and mice to reach the hive. Four wooden legs are not good either; when the hive is heavy and the ground wet, one or two may sink into the earth and cause the hive to tilt. Even if the legs rest on a brick or tile, I have known field mice to nest under the support, and then in the winter storms the tile subsides into a cavity and the three-legged hive is blown over. By far the best way is to sink two concrete blocks a couple of inches into the ground, parallel to each other and 2 ft 6 in apart at their centres, and rest two 5-ft angle irons across them, parallel and 14 in apart. Use a spirit level on the irons to get them absolutely level lengthways, and with a slight downwards slope from front to back crossways. One advantage of this type of stand is that a hive can easily be slid sideways without being lifted; very useful in some manipulations.

The stand shown in Fig. 9 would accommodate three hives for a short time, e.g. if a swarm were taken or a nucleus made, but is really designed for two. If you have a couple of paving slabs these could be used as hive stands for WBC hives, with legs, but a National or Langstroth hive should not be placed directly on the flat slab, as moisture would accumulate over the contact area. Put two wooden strips between hive and slab to allow air to circulate below the floor. It is convenient to place a flat sheet of polythene or a small piece of hardboard under and in front of the concrete and iron stand to prevent the growth of grass and weeds. Alternatively, a barrow-load of gravel or a spray round the hive with old engine oil or chlorate solution late in the evening will achieve the same purpose.

15 Large smoker with hive tool.

Hive tool and smoker

Very soon now you will be needing a smoker and a hive tool. I strongly recommend a Bingham smoker, with a bracket on the wooden backing of the bellows to take the hive tool, so that the two are always picked up together when needed. (See Fig. 15.) My favourite smoker cartridge is made by rolling some old sacking or hessian, plus dried rotten wood or fine wood shavings, in a 'Swiss roll' of corrugated cardboard, to fit the barrel of the smoker loosely, and twisting a strand of thin wire around to hold it in place (Fig. 16). Make up a weak solution of saltpetre (potassium nitrate), add some red ink and pour into a saucer; dip one end of the cartridge in it for a few seconds and allow it to dry for 24 hours. The red end will now smoulder easily if touched with a match, and an occasional puff of the bellows will keep it going all afternoon. Attach a cork or tapered wooden plug to the smoker by a length of wire, and cork the opening when you have finished. Without air it will go out and relight easily next time. You must leave it outside when you do this, as a safety precaution; there are beekeepers who have burnt down garages or set fire to cars with smokers left alight, and possibly knocked over and open.

Programme for May

a Look out for a couple of concrete blocks and two 5-ft lengths of angle iron;
b buy a smoker and a hive tool;
c make up half a dozen smoker cartridges;
d fix up your hive stand;

16 Smoker fuel cartridge.

e attend apiary meetings and talk to beekeepers.

f Your nucleus may be promised for the end of May but do not worry, you will not get it until early June. If you do, then read on. Practise handling frames and develop slow, deliberate hand movements, also the correct use of a hive tool, using an empty hive and frames.

Three

Early Summer – much ado

BEGINNERS – MAY/JUNE

First steps

The hive stand is in position and if you have not already acquired a hive, do so now: I suggest a National brood box complete with floor, entrance block, crown board and roof. This can be placed in position now. As you have ordered a four-frame nucleus you will need only seven additional frames, fitted with wax foundation, to make up a full brood box. Find out what sort of frames the nucleus will be on, whether Hoffman self-spacing or with metal end-spacers, so that your seven frames will match. The most satisfactory frames are the DN2 or DN5 with wide top bars and detachable wooden wedges: the narrower top bars invite more building of irregular comb. Both frames and wax foundation are normally sold in packs of ten, but a dealer will probably supply just the seven that you need. My personal preference is for diagonally wired foundation with loops at one end. Assembling frames and fitting sheets of wax foundation is a pleasant task, but if you have not done it before, don't treat it too lightly. It is in fact part of the test for the BBKA preliminary exam. A common mistake is to leave the wax sheet pushed into the grooves of the side bars but not fastened securely to the top bar; in the warm hive, with hundreds of bees clustered on to both sides of the wax, it slips down and crumples, so that one ends up with irregular-drawn comb, with a gap at the top and possibly joined to the next comb at a bulge half-way down. This wastes expensive foundation and involves more work for bees when rectified.

The frames will normally be supplied in the flat, and should first be assembled as shown in Fig. 17, the frame shoulders being held by gimp pins hammered in, and the twin bottom bars pinned at one end only until the wax sheet has been inserted, sliding into the grooves on the inner faces of the end bars. The loops of wire protruding from the bottom of the wax sheet should be bent over at right angles, and the wooden wedge strip replaced over the wires and secured firmly with two or three short sprigs, better pushed in by a ram-pin rather than hit with a hammer. Finally, the twin bottom bars should be lowered to engage in the slots of the other end bar, and fixed by another gimp pin.

The seven waxed frames should now be fitted into the brood box at each side, leaving a gap in the middle for the four frames of bees. If your hive is a National or Commercial (i.e. square in plan), you have the choice of

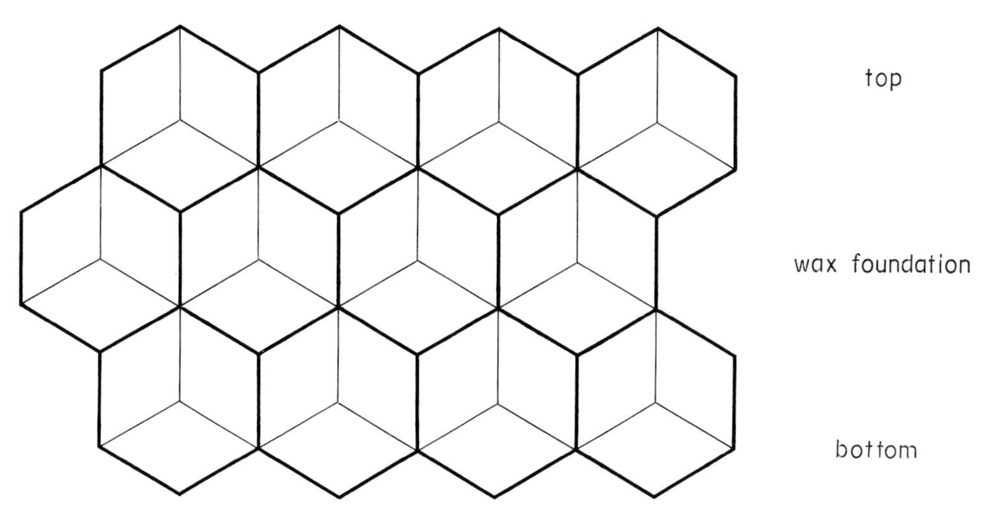

17 The right way up: wax foundation should always be fitted in frames this way up.
18 Fitting wax foundation.

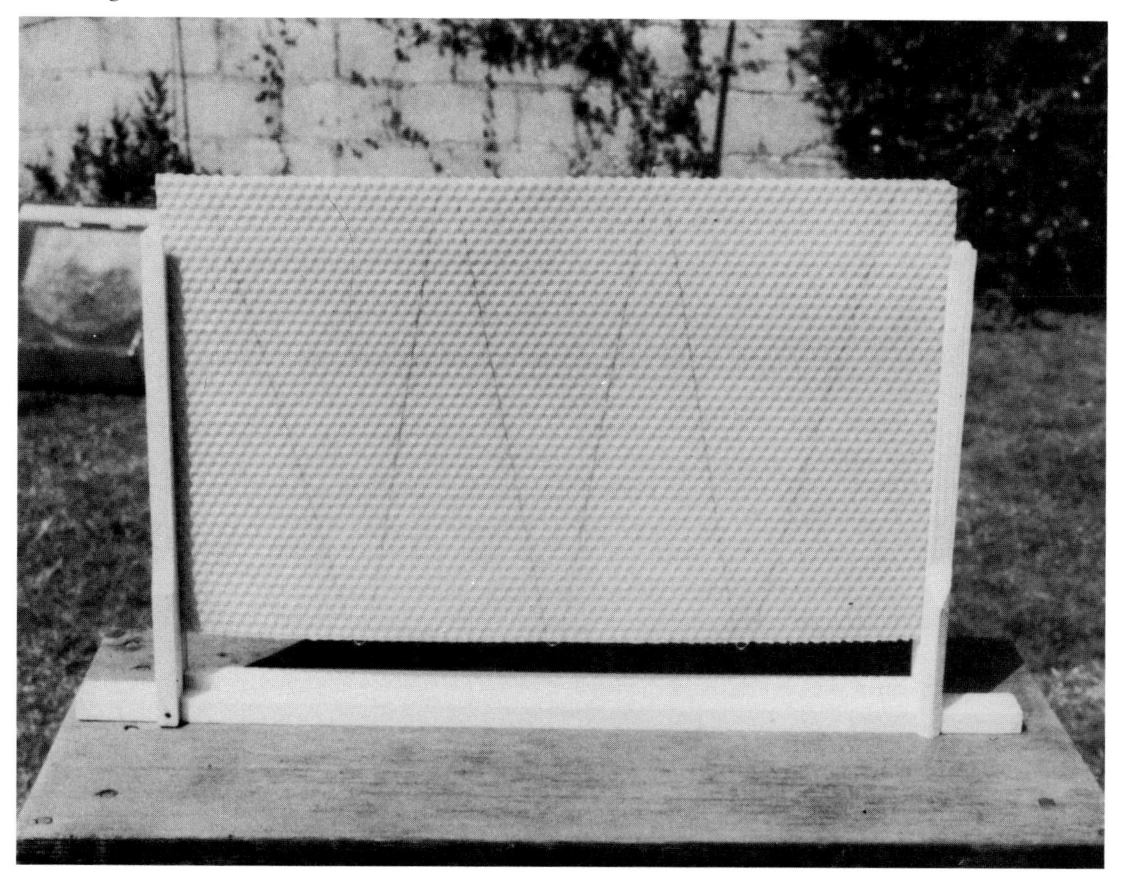

19 Frame assembled and fitted with foundation.

positioning the brood box on the floorboard so that the frames are either fore and aft (the 'cold' way) or square-rigged (the 'warm' way). It doesn't really matter, but the former is usually recommended as it provides parellel avenues leading from the entrance to the back of the hive, with more ready access to each comb, as well as simplifying ventilation problems. In the wild state bees will sometimes construct their combs one way and sometimes another – often diagonally.

When the nucleus at last arrives, move your hive a little to one side and stand the nucleus in its place; open the entrance and allow at least 15 minutes for the bees to fly freely. Now light up the smoker, direct two or three puffs well into the entrance and wait two minutes. Open up the top of the nuc. travelling box, giving a gentle puff of smoke across the top of the frames as you do so, and then transfer the frames one by one to your hive, keeping them in the same order as before. You may see the queen as you do this – take great care not to crush her, or any bees for that matter. It may help if you first take out one of the empty frames to give more room, replacing it at the side after pushing the others up to make firm contact with the four occupied ones. Now check that the queen is not among the few bees left in the travelling box, which should be placed two or three feet away from your hive, so that any bees still in it may fly out and join the rest. There will almost certainly be fanning bees at the hive entrance, heads down and tails up, with Nassanov scent gland exposed, giving the message, 'Come home – join us here.' Make up a gallon of sugar syrup by stirring eight pounds of sugar into six pints of warm water (roughly a kilogram to a litre). That evening, pour a quart into a round feeder placed over the hole in the crown board, and put an empty box or eke about three inches deep around it before replacing the roof. After a couple of days check the syrup level (by removing the hive roof very gently, not using smoke), and top up, doing this every three or four days if necessary. After allowing about five days for the bees to settle down, open up and see how they are doing.

Programme for May/June

- *a* Fit seven brood frames with wired foundation;
- *b* put hive on stand and check that it is complete;
- *c* buy a small circular aluminium feeder – a 2-pt (1-litre) size is large enough;
- *d* if you do not have a spare empty super (shallow box), make an eke by nailing together four wooden strips, each 3 in wide, to rest on the crown board and support the roof when the feeder is on;
- *e* transfer nucleus to your own hive;
- *f* feed steadily; inspect after four days, and then every four or five days, to gain experience and to watch the growth of the brood nest.

Hive manipulation

Plan to work from behind or to one side of a hive, never from in front, and keep clear of the flight path at all times. Get the smoker going well and give a gentle preliminary puff across the entrance to send any guard bees scuttling inside; then blow three firm puffs of smoke well into the entrance, one straight in and one to each side. You should already be wearing a bee-suit and wellington boots with trouser legs tucked in, so while the bees are reacting to the smoke you have two minutes to put on your veil and check that it is well tucked in at the neck; also gloves, if you are a beginner. Bees have been forest dwellers for millions of years, and the smell of burning wood evokes the age-old response, 'Fill up honey sacs and stand by to evacuate if it gets too hot.' Work by bee researchers at Rothamsted some years ago showed that it takes about two minutes for this response to become effective, and to override the normal response of 'Scramble – repel intruder' when a hive is knocked or opened up. In a year or two, when you have gained much more experience, you may be able to sense when the bees are so busy and happy with nectar pouring in on a mild day, that it is possible to open up very gently without smoke, talking quietly to the bees and working with slow and deliberate movements so that the bees seem not to notice. There is, I think, something one might call the 'smell of fear' which horses, dogs and bees can detect and react against. Possibly this arises from adrenalin secreted by our bodies when we sense danger, and if one is worried this effect is more pronounced. Even so, one should always have smoker and veil ready to hand in case the mood changes.

Push the hive tool gently between the crown board and brood box *at a corner*, lever up and puff smoke into the narrow gap before slowly lifting up the crown board. Look closely at the bees on the board – just once in fifty times the queen may be there – and then lean the board against the side of the hive and replace it with your inspection cloth. Roll this back to expose the tops of two frames at one end; open up a slight gap between them, using the hive tool as a lever, and gently lift up the end frame. In spring and early summer this may contain nothing but sealed stores with few bees on it, and after a quick inspection it may be placed on the hive stand leaning against the side of the hive. This leaves a gap, making it easy to lever each successive frame away from its neighbour before lifting it out for inspection and thus avoiding rolling bees against a comb face, which they dislike.

Hold the frame by the lugs with both hands, above the hive, and check what there is to be seen – on this second frame it may be just a sheet of foundation with a dozen bees on it, or heavy with food. Now drop your right hand until the top bar is vertical, rotate the frame through 180° and bring your right hand up again, so that you are now looking at the opposite side of the comb, without the face of the comb having been horizontal at any time. This is important, as horizontal combs can drip nectar, and may sag in hot weather. Finally, go into reverse and replace the frame, pushing it firmly against the side of the hive to leave a gap before the next frame. Carry on like this, rolling the cloth back a couple of inches at a time, and work your way right through the box, finally replacing the first comb at the far end, but keeping all other frames in the same relative positions. If you forget which way round a frame should be replaced, remember that the end with more honey came from the back of the hive, away from the entrance.

Use slow, deliberate movements, especially when your hand passes over the open hive; give a puff of smoke if the bees seem to be coming up in any number. When half-way through the frames, switch the cloth to cover frames already inspected and roll it out to cover each successive frame replaced. Watch for the first comb face with a large patch of pollen – this tells you that there will most likely be eggs or brood on the next comb. Look for and identify sealed brood, open brood and eggs. It helps if you hold the comb so that the light

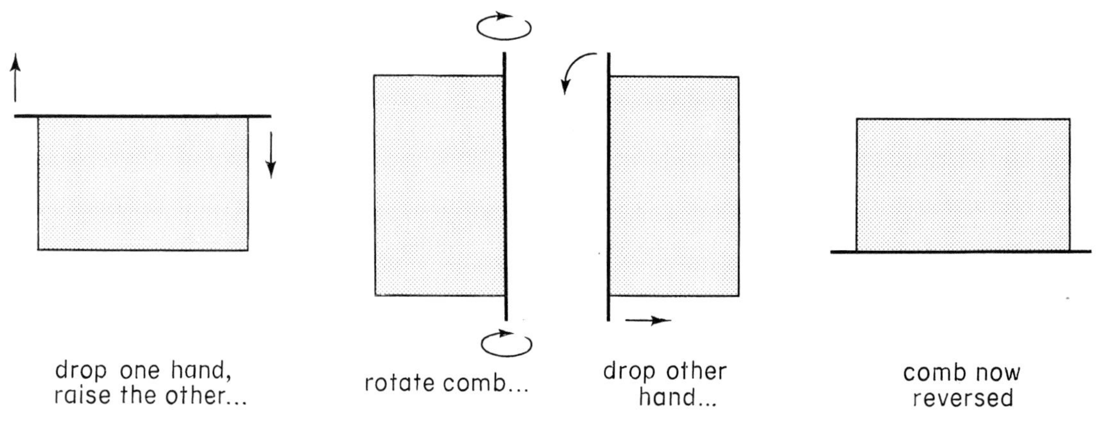

20 Comb inspection: the comb should be kept vertical at all times.

shines over your shoulder into the cells; eggs in particular are very much easier to spot if you are fortunate enough to have the sun shining down into the base of the cells, not shaded by your hat! If bees are clustering tightly and obscuring the comb, press very gently with the back of the hand to disperse them. Queen-finding is the subject of another section, but in general terms she is most likely to be found on a comb of eggs, and least likely to be found on a comb solid with food.

Even in the central combs there should be honey at the corners and in a narrow strip of comb at the top; in the combs at each end of the brood nest there should be much more than this. If not, then you must feed, even at the end of June. Notice how some bees are clinging to each other in a loose cluster at the edge of the occupied area, or even hanging in festoons from the frames: these are the wax builders, drawing out new comb. It is unlikely that there will be much, if any, drone comb in your nucleus, but it is easily recognized by its larger size and domed cappings.

Beginners sometimes confuse sealed brood with sealed stores, but remember that brood is more likely to be placed centrally and sealed stores at the top and sides. Sealed brood has the colour and texture of a wholemeal biscuit, while sealed honey has much lighter cappings, often with a ribbed surface. Once you have seen both on the same comb, you will never be in any doubt again. You may spot a young bee biting its way out of a capped cell, and some very young bees on the comb, still downy and damp from recent emergence.

It is best neither to hurry unduly nor be too slow, but never waste time when a hive is open. Every manipulation should have a purpose, and if it is a general examination in April for the first time that year, have at hand one or two spare frames fitted with foundation, to replace any old, gnarled or misshapen black combs. As a beginner the purpose of manipulation may just be to gain experience, and this is important, not only for you but also for your bees. Before replacing the crown board, a puff of smoke may be necessary around the top edges of the box to get the bees down and avoid crushing them. If you have a record card on the crown board, make a brief note of date and condition and replace the roof.

Having tried oil of wintergreen, weak carbolic solution, beeswax and other substances rubbed on hand before manipulation, I have found that a smear of oil of cloves, well rubbed into hands and wrists, is the best sting-deterrent, but an occasional sting will still come one's way.

IMPROVERS AND EXPERIENCED BEEKEEPERS

Checking hives on cold days in May or June

The dilemma is that after perhaps two warm weeks with nectar pouring into hives, the weather pattern changes and there is a cold spell lasting a week or more. One sets out hopefully but on arrival at an out-apiary fifteen miles away the sky is cloudy, there is a cold north-east breeze and the air temperature even at midday is just under 50°F (10°C), although the weather man said it would rise to 58–60°F (15°C). Scarcely any bees are flying at noon, and when the first hive is opened up they obviously resent the intrusion. What should one do?

One solution is to carry out a limited inspection of two or three of the strongest stocks (out of 10–18, perhaps), and if they are not showing any signs of swarming then pack up and go. Proceed as follows with a single brood chamber. After smoking the entrance, wait two minutes, remove and cover supers placed diagonally on the upturned roof, and go directly to the centre of the brood nest. Ease a central comb by levering with a hive tool on both sides and lift it out, placing it in an empty nuc. box and covering with a cloth. Look down at both exposed comb faces and lift out a second comb and check for queen cells. Replace and pull to one side to allow access to another comb; inspect this too, checking the newly exposed comb face with a quick glance. Finally, check and replace the first comb in its original position and close up. If there were no queen cells, then the odds are that no swarming is likely for another ten days. If one or two 'dry' queen cups are seen, i.e. queen cups not yet laid in, then one is reasonably safe. One can never be 100% certain of anything with bees, but nine times out of ten if no swarming preparations are seen on the central combs of the strongest hives, one is justified in leaving the matter there. This whole process takes only minutes, with no harm to the bees. Those who run hives on a double brood chamber (or one and a half) can get an indication even more easily by levering up the top brood box, pulling it forward an inch and tilting so that one may see underneath; queen cells are most likely to be visible under the top box, if there at all. When there are supers on the hive this is not so easy, but it can be done, preferably with a second pair of hands on the pile of supers while the first person gets down and looks in.

If there are queen cells then a second visit must be made very quickly, or else some anti-swarming measures taken on the spot, despite the cold day.

Supering

Some books say that the first super should be put on in May, when the bees are covering all but one of the frames in the brood chamber, and the second when the bees tell you to, by depositing white new wax on the crown board. In the south of England it is necessary to have the first super on before the middle of April, if not on all winter. As to the rest, if the bees are putting new wax at the top, they are not saying that it is time for the next super now but that you should have put it on ten days ago. Later supering can mean that if a sudden flush of nectar comes about mid-April, the only empty cells are those in the brood chamber that the queen needs for egg-laying. This causes a congested brood nest and a reduction in the larvae just when the colony should be expanding most rapidly. It is far better to super ahead of the bees' needs, and if you are worried that an empty super creates a large, cold space which discourages the bees, then put a single sheet of newspaper under the second super to stop heat loss by upward convection currents. The bees will soon chew it away when they are ready to expand, and the fluffy chewed-up paper coming out of the entrance will tell you at the time.

If you have an out-apiary many miles from home which you can only visit once in a while, a good method of supering up in one go is to put on three at once, with a single sheet of newspaper between the first and second, and a sheet of very thin card (or three sheets of newspaper) between second and third. I thought that I had invented this system, but discovered recently that some beekeepers did it during the Second World War, when petrol was severely rationed. Try it just for interest on a hive in the garden.

There is also debate on whether to add additional supers above or below the first. In my experience little is gained by complicating the procedure, and straightforward top-supering is as good a method as any, except perhaps for getting new foundation drawn out on close-spaced frames. If working for comb honey, then a clearer-board put under a super of drawn comb even before it is completely full will help by crowding the bees into a box of sections or comb starters placed under it, but do not keep this box close to the brood nest more than two or three weeks or pollen may be deposited in central combs, and the clean white cappings discoloured by thousands of little bee feet.

One good reason for having two, or preferably three, supers on a hive before the end of June is to ensure that there is enough room for the bees, quite apart from the honey which may or may not be produced. In a poor year this may result in honey packed only in the central five to seven super combs, but if swarming has been reduced it will have been worth while.

Normally the main honey flow tapers off towards the end of July, and it may pay then to rearrange combs, changing over two central combs full of honey with the two flank combs possibly almost empty. This saves unnecessary super handling when the crop is taken off and extracted. This same 'ends-to-middle' technique can also be used if on an apiary visit earlier in summer one is caught without spare supers, but within a few days the next one must go on. Conversely, if one takes supers to an out-apiary and they are not needed immediately, put one on each hive over the crown board (feed hole open). When ready to go up, the bees will do so, and on the next visit the crown board can be moved above.

In areas of oil-seed rape a different approach is necessary, as honey left on will granulate in the comb and be difficult or almost impossible to handle. Here the management technique is to clear supers two-thirds capped into an empty super placed below, and to extract the full supers as soon as they are taken off, keeping them warm in the meantime.

It is easy to write about taking off a super, but in practice they are often stuck together with wild comb built in between and are difficult to handle. The problem arises because frame bottom bars are normally parallel and close to the frame top bars immediately below; the bees treat two (or even three) frames so aligned as part of a long, continuous comb such as they might build in their wild state in a hollow tree. I found out quite accidentally some years ago that a super placed with frames at right angles to those in the super below does not get stuck to anything like the same extent, and in 1983, when up to five supers had to be taken off the best hives, I encountered no problem with supers so placed. Orthodox beekeepers may say that this method of supering is incorrect, and that the correct bee-space between the supers should prevent the problem. This view is based on a fallacy. Bees respect a bee-space around the sides of frames and, when it suits them, even over the long lugs of correctly spaced BS frames, but parallel honey combs placed above each other are seen as *continuous* structures, and the gaps bridged with wild comb three times out of four.

WORKING FOR SPRING HONEY – IMPROVERS

Honey flows

Although bees may be gathering some nectar from March to November, there are usually only short periods of time when large quantities are coming in, and these we call 'honey flows'. To secure a surplus for the beekeeper, three conditions have to be satisfied; *a* large numbers of nectar-producing flowers in full bloom; *b* large numbers of adult, flying bees in the hive; *c* calm, warm weather, not too dry.

Usually there are three possibilities: the early spring flow between the third week in April and the end of May from oil-seed rape, fruit trees and bushes, sycamore and hawthorn; the main flow from lime, clover, blackberry, and willow herb (mid-June to end of July); and the heather flow from ling on the moors in August. It is possible to get a surplus from any of these flows if conditions are right, depending on the weather, the locality and the strength of the bees, but usually in Britain only one or at the most two of these flows will oblige in any particular year.

Here are two suggestions for getting a surplus from the spring flow.

(A) 'Shook' swarm

Choose the strongest colony and feed it a gallon of thick syrup between 10 and 20 March. Prepare a shallow super of frames fitted with wax foundation, and have ready a spare hive floor, crown board and roof. About noon one warm day, about the third or fourth week in April, when the dandelions are coming up to full flower, put the spare floor 2–3 ft from the hive and pointing away from it, either behind or to one side. Give two or three puffs of smoke, wait two minutes and lift the brood box off the hive on to the spare floor, then replace it with the new shallow box of foundation, for the flying bees to return to and occupy. Now open up the hive, take out the brood frames one by one and shake the bees into or on to a cloth just in front of the new box. Check each comb for the queen, and if she is seen place the comb temporarily in the inverted roof and cover with a cloth.

Transfer of bees

Carry on until about two-thirds of all the bees have been transferred, either shaking or brushing them off, but leave enough bees to cover the brood in the original box. If there are more than five frames of brood, give some (without bees) to weaker colonies or to nuclei which have come through the winter. Leave a comb of eggs or very young larvae from which a queen may be raised. Coming back to the comb with the queen, gently push on a queen marking cage to restrain her, and with the thin blades of a pair of fine scissors clip off about one-third of a wing on one side. Then gently brush her, with some bees, into the new box. Put on a queen-excluder, followed by the super of drawn comb (plus bees) from the old stock, then a super of very thin foundation for comb honey, and finally crown board and roof. If weather conditions are right, the 'shook' swarm will draw comb and store nectar, for some days, as it has a large force of bees and no brood to care for. They may try to swarm out during the first 48 hours, but the clipped queen will be unable to fly. Mostly they don't, so if the queen is not found, carry on just the same.

Move old hive

After a few days, move the old hive back alongside, with the entrance pointing in the same way, and later (two weeks after making the shook swarm) move it away again, thus shooting more flying bees into the 'swarm'. By the end of May, if the weather is kind, it should be possible to take one or even two supers of spring honey. After checking that the old hive has raised a young queen, the brood box can be brought back and united to the swarm, putting the shallow box of brood over a sheet of newspaper on the original deep box. The young queen will take over (nine times out of ten), and the combined stock is well placed for taking advantage of the main flow, from mid-June onwards. One final word – there is no point at all in making a shook swarm from a weak stock. Unless the brood box is crowded with bees, with at least four or five frames of bees in a super (or half-brood box of a one and a half hive), there will not be enough workers to gain a surplus.

(B) Concentration, or 'plumping'

Arrange matters so that you have two strong stocks next to each other and feed both in March. On a warm day towards the end of April put a queen-excluder and two supers on the stronger hive of the two, and move the other two or three feet away, turning the entrance to the rear. Move the stronger hive to the old position of the moved hive; this will receive several thousand flying bees from the other hive. After a week bring that hive back again near to its original position and seven days later move it away again, to give a second reinforcement. On the principle that one very strong stock will store more than two medium ones, you may get a super or two of spring honey, especially in a suburban area or near fields of early oil-seed rape. You may also get a large swarm from the strong hive! If you do, hive it on a shallow box of foundation on the original site and proceed as in method (A), supering up the swarm and moving the parent stock to one side. As a variation, one can make a shook swarm from the stronger hive after the first reinforcement, thus anticipating any natural swarming.

Both these methods have worked well for me in two out of the last four years, when a good spring flow was followed by a poor summer. One the other hand, neither method was practicable in 1983, when the weather stayed cool and wet until well into June, and many colonies needed feeding in May.

Oil-seed rape

It is probable that this crop has now taken over from clover as the major source of nectar in Britain, and the acreage farmed is still increasing. As one drives across the country in summer, especially in the Midlands and East Anglia, huge fields of these bright-yellow flowers are seen everywhere. How has this come about, in the space of ten or twelve years? It is largely on account of the EEC policy of becoming self-sufficient in the production of vegetable oil, used in the manufacture of margarine, salad and cooking oils, and also as a specialized lubricant. A subsidy of nearly £150 a ton plus the possibility of a good 'break' crop offers many advantages to farmers.

Although differing varieties, some winter-sown and some spring-sown, flower at different times, so that somewhere or other there are bright yellow fields of rape over a good deal of the summer, it is generally speaking an early crop, flowering from late April to the end of May. This raises problems immediately, as stocks have to be strong much earlier in the year. However, the main problem is the tendency of rape honey to granulate within days, even in the combs before extraction. This is because of its high glucose content (with less fructose than normal honey).

Management techniques

If the beekeeper merely wishes to get combs drawn and build up stocks for other crops later in the summer, no special management is required, except perhaps to add a second brood box of foundation to be drawn and stored for subsequent use in making increase or nuclei for sale. For honey production the stocks must be strong earlier than usual, which calls for stimulative feeding in mid-March and again in April – at least a gallon of thick syrup on both occasions. In some districts it may also be necessary to feed a cake of pollen substitute. The technique described on page 43 ('Working for spring honey') would apply.

The main problem arises from the rapid granulation, and it is now common practice to remove supers and extract as soon as the central combs are partly sealed. So long as the open combs do not drop honey when held horizontally over the hive, they may be extracted.

Early summer – MUCH ADO

Rape honey should not be left overnight in the extractor but run into 28-lb tins or plastic buckets immediately, and these subsequently warmed up as described on page 169, before being strained and bottled. If there is an immediate need for honey to sell, then straining and bottling must be done on the same day, or crystals already forming will block the straining cloth. The honey is white and so finely grained as almost to resemble lard; it is very sweet but thought by some to be lacking in flavour and character, so that it is often stockpiled and later blended with summer floral honeys of a darker colour and stronger flavour.

Another technique is to 'cream' rape honey, using the method described on page 170.

Spray dangers

There are insect pests which can infest oil-seed rape, and some of the insecticide sprays used (like triazophos and HCH) are very toxic to bees. For example, ADAS suggest spraying against pollen beetles when a count reveals three or more per plant on spring-sown rape and 15–20 or more on winter rape. Sprays are also recommended against seed weevil and pod midge when an average of more than two per plant is found, but this is usually at a late yellow bud stage when bees would not normally be working the plants. The difficulty arises when insect pests are detected late, or when spraying contractors cannot carry out orders for some days, and by the time they arrive the plant is in flower and being visited by bees. In Britain, treatment of rape flowers with insecticide is not forbidden, although ADAS ask that it should not be done. Some European countries have stronger policies and specifically forbid the use on open flowers of insecticides poisonous to bees. The closest possible liaison between farmer and beekeeper is therefore much more important in Britain.

FINDING THE QUEEN – BEGINNERS AND IMPROVERS

Why find her?

Some beekeepers say that they have never seen a queen, yet still get good yields of honey. However, if an efficient swarm control system is to be practised, or an artificial swarm made in spring to replace a lost stock, or a nucleus made for a friend, the queen simply has to be found. Many successful beekeepers re-queen automatically every other year, and realize, as Brother Adam says, that the queen is the mainspring of the life of a colony, and that by replacing her at will we have the power to rejuvenate and keep a colony perpetually young and at its maximum productive ability. There is also much more satisfaction in acquiring the basic skills and technique of the craft than in just owning a couple of hives and hoping for a good season.

When to find her

The best time of year for queen-finding in an over-wintered stock is April, before the hives get really crowded and when there is a large proportion of young (and gentle) bees. The best time is between 11 a.m. and 3 p.m. on a fine day when more bees are likely to be flying, leaving fewer bees on the combs. A regular practice of queen-marking in April may save a lot of trouble later on. If making nuclei on just three or four frames, one might as well mark the young queen with the colour of the year *as soon as she has started to lay*, whether in May, June, July or even August.

Normal approach

Use a minimum of smoke in the entrance, wait a couple of minutes and open up as stealthily as possible. Use only a drift of smoke across the tops of the frames as the crown board is taken off, and cover up immediately to prevent light

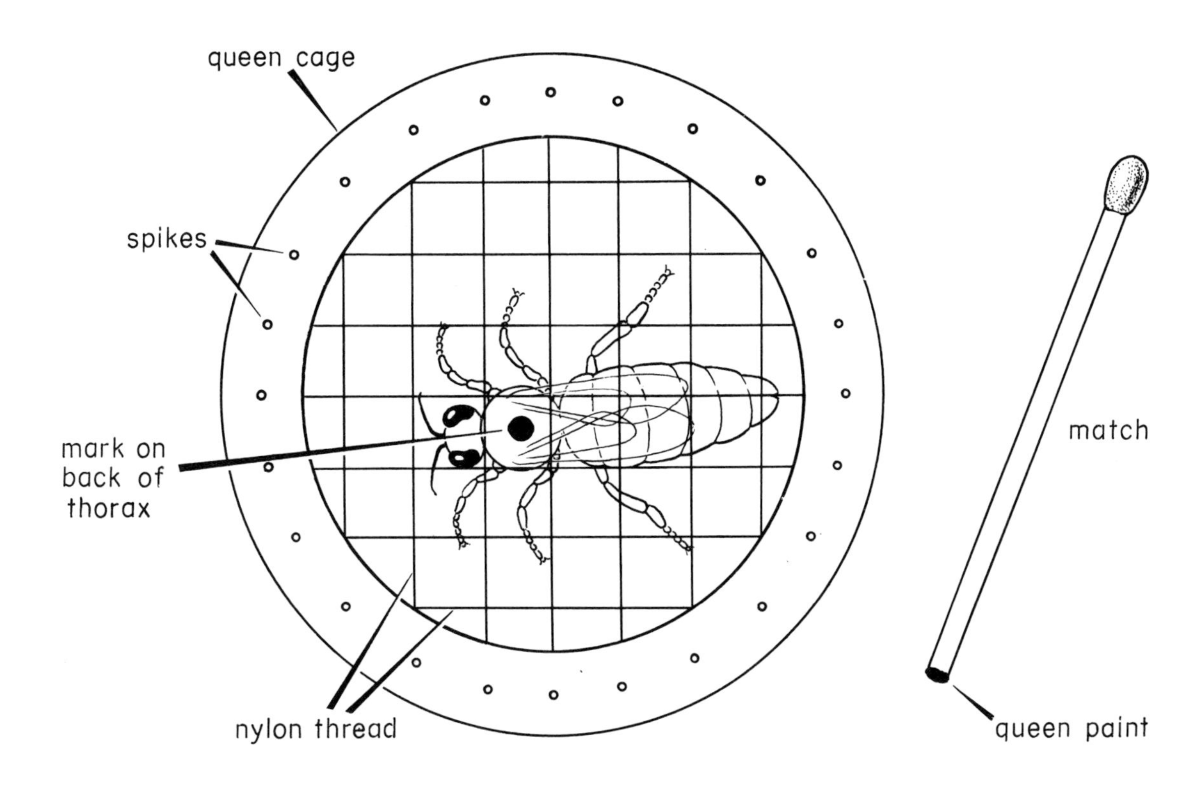

21 Marking a queen. International colours for year ending:
5 or 0 – blue
6 or 1 – white
7 or 2 – yellow
8 or 3 – red
9 or 4 – green

Lower the cage gently over the queen on a comb, wait until worker bees have escaped between the spikes, and press down gently until queen's thorax is framed in a nylon square. Dip match end in queen paint, press on woodwork to remove excess, and dab gently on queen's back, rotating match between finger and thumb to give a circular mark. Wait a few seconds, then lift cage quickly to prevent nylon thread dragging across soft paint as queen struggles to move. Special queen paint can be obtained from bee dealers, or any quick-drying paint or lacquer may be used, apart from those based on amyl acetate (cellulose 'thinners').

driving the queen down. Check the under-surface of the crown board to make sure the queen is not there. If a queen-excluder is on, then remove very gently and check this instead. Roll back the cloth two inches and lift out the end frame on the near side, check for queen and place in upturned roof or another box. Take out the second frame also, and if no brood is seen, place alongside the first, to give more room for seeing down into the hive. Gently loosen the third frame, pull it away an inch and lift; as it comes up, glance down at the freshly exposed face of the next frame, and you may see the queen walking downwards away from the light. Now raise the frame to eye-level, with the light behind you, and look first along the bottom edge, then round the sides and top and spiral into the centre. Reverse the

22 Queen-marking cage in use.

comb and repeat. Finally defocus the eyes, as if looking at a distant object, and look at the entire face of the comb as you lower it and pull it firmly against the side of the hive. Look for six long legs with light amber stockings, rather than for the queen all the time. With slightly defocused eyes and a frame held at arm's length, one is more likely to notice something different. Work through all the brood combs in this way, paying most attention to combs with eggs, very young brood and empty cells, and least attention to combs solid with sealed brood, or pollen and honey. In this way the first inspection should only take four or five minutes and (with practice) the queen will be found, three times out of four.

Second time through

If the queen is not found, one can close up and hope for better luck next time, or press on and go through the frames again. Repeat the procedure, pushing back each comb to its original position as you go. This time pay special attention to the space (if any) between comb and bottom bar of frame, and any gaps at the sides or fissures in the comb, clearing out bees with a puff of smoke. If there is a knot of bees on any comb, gently lay on the backs of two fingers to disperse them. Last year I found two queens (both dark) on the same day in two successive hives, playing 'dodgems' on the bottom bar and moving from one side to the other as the frame was reversed. Once, when making artificial swarms with strong stocks at an out-apiary 14 miles away (so the job had to be done in one visit), I missed one queen twice and only saw her on the third time through because an egg was still protruding from her abdomen, the only part visible on a very crowded comb. On the second time through, watch the bees on the tops of the frames closely. You may see them fanning or running in one direction, towards the comb with the queen on.

What next?
If the queen is not found on the second time through it might be as well to try again another day and move on to the next hive. If she just has to be found for some very good reason, then other methods may be tried. One useful technique is to remove four or six central combs to another brood box on a floor (or upturned roof) and place them in pairs with two or three inches between each pair. Do the same in the original box with the remaining frames and cover with cloths for a few minutes, then look quickly on the inside faces of each pair of combs, where the queen is most likely to be.

If the purpose of finding the queen is to make an artificial swarm, or start a two-queen colony, then move the parent stock a few feet away and collect the flying bees in a new brood box with combs or foundation placed on the old site. As flying bees return to this, the parent stock becomes less crowded and the queen easier to find.

Other points
Perhaps one queen in 20 may be a 'bottom-runner' and be found on the floor board in a corner; excessive smoking on the frames may also drive a queen down to the floor. Generally speaking, if very little smoke is used at the beginning then the queen may often be seen going about her normal business of looking for empty cells and laying eggs in them, with the orthodox 'court' of bees facing her in a rough circle.

If double brood chambers are involved, then a queen-excluder may be slipped between them a few days before, and from the state of the combs in the top box (presence or absence of eggs) one at least knows in which box to look for the queen. Working with a large commercial bee business in New Zealand one year, where no excluders were used throughout the summer and only pressure of stored honey kept the queen down in the bottom two boxes (Langstroths), we cleared the honey supers down to the bottom two boxes with benzaldehyde cloths and slipped a queen-excluder between the remaining boxes after taking the honey off. This was for the sole purpose of saving the re-queening crews the trouble of looking through both boxes a week later.

If the purpose of finding the queen is to take her out and introduce another one, beware of the possibility of a 'mother/daughter situation' after a supersedure, especially late in the season. Check for the remains of an emerged supersedure queen cell (usually on the face of a central comb) before putting in a new £10 queen, or you may find her dead on the doorstep next morning, as the bees prefer a young queen of their own raising.

QUEEN-REARING – ALL BEEKEEPERS

In order to produce first-class queens, capable of heading strong colonies of good bees for at least a couple of years, two main requirements have to be satisfied.

a The virgin queen must be of good parentage, have been generously and continuously fed as a larva, and have had extra royal jelly packed into her cell before it was sealed. (Unlike a worker larva, she continues to feed for some hours after her cell is capped.) She must also be well nourished during those few days before she goes out on her mating flight or flights.

b Before she is three weeks old she must have at least one warm day (60–65°F or warmer) on which to mate on the wing with up to six or seven drones, also of good parentage, so that her spermatheca may receive a full charge of sperms to last her through her working life. The drones are quite likely to have come from her own hive, but may well come from any hive up to six miles away.

Obviously we have no control over the weather, and in practice not usually much control over which drones she will mate with, but at least we know her mother and can arrange for the rearing colony to be prosperous, and well stocked with nurse bees. We may also arrange to flood a mating apiary with suitable drones, by putting a comb of drone cells in April into a chosen hive. It is possible to inseminate artificially, or to maintain isolated apiaries on small islands where only the best drones are permitted, but this will be outside the scope of most beekeepers. In nature, queen cells are built in three different situations:

Swarming
In summer, usually between mid-May and the end of June when a colony is crowded with young bees and has a comfortable surplus of food, swarm cells may be built so that when a swarm leaves with the old queen there are several princesses left behind, one of which will mate and take over the original stock. Swarm cells are usually made on the outer edges of brood combs.

Emergency
If a queen is accidentally lost, for example when a hive is knocked over by an animal, or the nest in a hollow tree damaged by a storm, the bees will make queen cells by widening and lengthening a few existing worker cells containing very young larvae, and feeding them with royal jelly. This can happen any time between March and September. These cells may be made almost anywhere on the combs.

Supersedure
When the bees realize that a queen is failing, or is in some way damaged, they may produce two or three queen cells (sometimes only one), usually towards the centre of a comb in the middle of the brood nest.

As beekeepers we can make use of queen cells produced under any of these three impulses, always bearing in mind that good nursing calls for a large number of young bees. These must be well fed on a diet of both pollen and honey so that their glands may secrete ample amounts of creamy, protein-rich brood food.

In the absence of a honey flow, keep a feeder of syrup on the top box, and make sure a good comb of pollen is there too; in some areas a cake of pollen substitute may be necessary.

Swarm cells are produced naturally under good conditions, so that the emerging virgin is usually large, active and very satisfactory, so long as no undesirable features have been inherited from her parents. The original crop of swarm cells is best, and if these were broken down as a swarm control measure and a second batch built, the replacements are often more numerous, less well provisioned and generally inferior.

When a queen is removed from a prosperous colony in May or early June, the *emergency cells* then constructed may be as good as swarm cells, so long as the basic requirements are satisfied. However, if a nucleus is made by taking four combs from a colony and the bees are left to raise their own queen with a labour force often inadequate and having many other mouths to feed, an inferior queen may be expected. Young queens produced by supersedure can be very good, and there are the great practical advantages that no bees are lost from the work force by swarming, and that there has been no check in egg-laying. Supersedure queens (outside the main swarming season) are also more liable to mate with brother drones, so that there is the likelihood of perpetuating the existing good (or bad) characteristics. There is also the disadvantage of inbreeding, which may cause an increase in non-viable brood (more pop-holes in area of sealed brood). In practice supersedure is probably to be welcomed over a five- to seven-year period, after which new blood is necessary either by natural swarming or queen introduction.

MAKING NUCLEI AND RAISING QUEENS – IMPROVERS (JUNE)

Although new beekeepers may have read books and attended lectures on that favourite subject, swarm control, in reality most people find great difficulty in searching through a powerful stock of bees in June hunting for a queen or cutting out queen cells. Not for a couple of seasons will the average person be competent enough, or have sufficient confidence, to do this. In practice it is almost inevitable that one day in late May, June or early July the bees from one stock in every three or four will swarm. This can be the occasion for the simplest of all methods of nucleus production and queen-raising.

The actual swarm, once taken, should be removed to a shady spot to rest until 5 p.m. or later, and then hived as already described in a new box containing frames of drawn comb or foundation, placed on the old site. The swarmed stock should be moved two or three feet to one side and turned away so that the entrance is facing in a different direction. The queen-excluder goes on the new box, plus supers taken from the swarmed stock. By the next day the flying bees from the old stock will have reinforced the swarm, and the old stock will be very much more docile as well as less crowded, having a large proportion of very young bees. This means that even a beginner should be able to go through each frame and assess the numbers of queen cells, the frames with brood and the amount of pollen and honey present.

Making nuclei

Now is the time to make two four-frame nuclei and leave three combs in the centre of the old box, with four frames of wax foundation on each side. First choose a comb with at least one queen cell and place it in a nucleus box, also another comb of sealed brood, and a comb of food on either side, plus the bees clustering on these combs. Repeat with the second nucleus box and check that among the three combs left in the old hive there is at least one queen cell. Usually swarm cells are distributed on at least three different combs, but if not, then one or two should be carefully cut out with a penknife and gently pushed down between two frames. At this stage it does not matter if a comb is damaged by cutting out a cell, as the bees will rapidly repair the hole made; what is important is that one should cut at least a half inch all around the queen cell to avoid damaging it. The two nuclei should be moved away from the main hive and left to themselves for the next two or three weeks. If

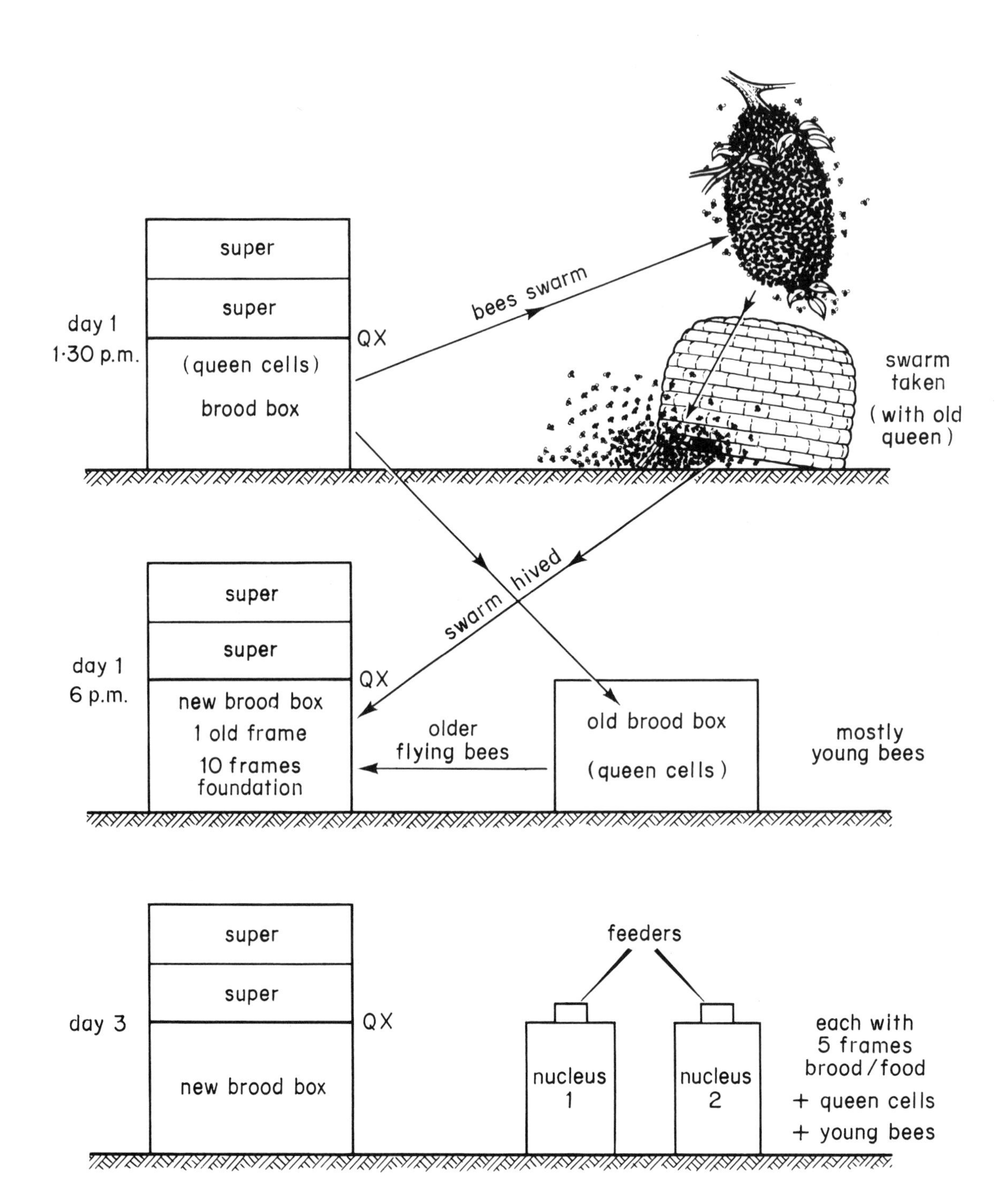

23 Making nuclei.

in the same apiary, stuff grass or tissue paper in the entrance for the first 24 hours, to cut down loss of flying bees to old hive. At first there will be very little activity as they will retain only young bees, which will stay at home to incubate the brood nest and feed any open larvae. After about a week there will be a noticeable increase in bees seen flying, and after three weeks all the brood will have emerged and the nucleus should be flying freely. If all has gone well the new queen will have mated and will be laying by this time; heavy loads of pollen will then be seen going into the nuc. and the bees will be flying with a new sense of purpose and energy. This is the time to open up and check that brood is present; eggs can be difficult to spot sometimes but a patch of pearly-white larvae is difficult to miss.

The three-frame nucleus left in the original box will have had the advantage of rather more bees than either of the two four-frame nucs, and will soon catch up as the new queen begins to lay. The stock in the old box will need to be fed unless it was left at least one good comb of food, but no feeding at all must be done for the first 48 hours, as there is the possibility of a few of the older bees drifting back to the swarm and carrying the message that a social security hive is close by, with resultant robbing. In any case, the stock with wax foundation must be fed continuously after a few days, as wax foundation has to be drawn into comb, and this stock has to build up, as must the two nucs, of course. In fact all three nucs will need a great deal of sugar syrup for several weeks. After the honey crop has been taken off the swarm (end of August or early September) it would be good practice to unite the old box with its young queen back to the swarm.

Using spare queen cells

Normally there will be at least eight to twelve queen cells in the swarmed stock, and so far we have only needed three, although it would not matter if each new unit had more than one. Some beekeepers in New Zealand have developed a method of re-queening with queen cells without having first to find and kill the old queen, so that spare sealed queen cells might well be cut out and gently pushed down between the middle top bars of any colony with a queen believed to be two or more years old. Even with no protection, such an introduced cell will probably be accepted as a 'supersedure' cell from which a young queen will be reared to take over from the old lady without swarming. The chances of acceptance are higher if the sides of the cell are protected while the tip is left free. The reason for this is that when queen cells are rejected and torn down, they are always attacked at the side, never at the tip, so that with a protected cell the virgin will at least be able to emerge undamaged in a simulated mother/daughter situation and have an 80% chance of acceptance. The standard method of cell protection in this country is by a spiral spring or a short section of hose pipe, but I have found a 2-in (5-cm) length of normal adhesive tape such as Sellotape wrapped round the sides to be equally effective.

24 Protecting the queen cell.

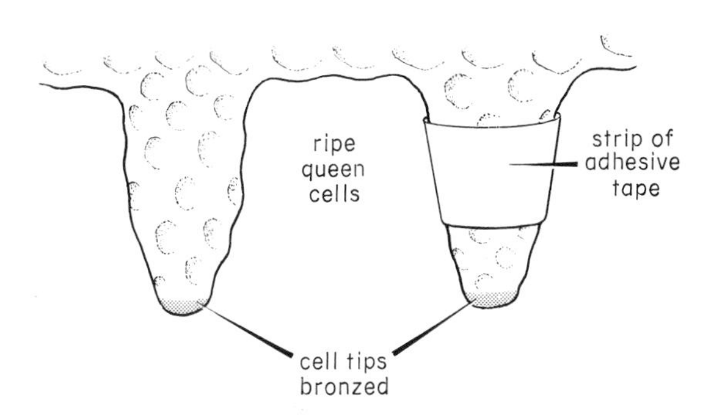

Gentle handling

As a larva, the young queen is held on her bed of royal jelly by surface-tension forces only (i.e. by natural 'stickiness'), and when her weight becomes appreciable (the third day after hatching), any shaking or jarring of the comb can 'shake her out of bed' and separate her from the reservoir of royal jelly on which she should be floating. This can sometimes result in an imperfectly developed queen in an unusually long queen cell. Even after sealing and pupation the young queen is still vulnerable, as any pressure on her developing ovarioles (egg tubes) can cause damage. Her ultimate efficiency as an egg-layer will depend largely on the number and quality of her ovarioles, which make up the huge twin ovaries housed in her large abdomen. A good queen should have at least 340 perfectly developed egg tubes. To prevent even the pressure of her own weight affecting them (if lying on her side), nature has arranged that queen cells (unlike those of drones and workers) hang vertically. Only in the last two days before natural emergence does the chitin of her abdominal wall harden sufficiently to withstand even light pressure. Hence the importance of gentle handling. At Buckfast Abbey, Brother Adam transports queen cells to the mating apiary in a special carrier, over a warm water tank placed on a three-inch deep foam rubber mat, in a vehicle with tyre pressure kept as low as possible to avoid any jolts and vibrations. On a much smaller scale, a few cells at a time may be safely stored for twenty-four hours and delivered to an out-apiary in a large Thermos jar, as shown

25 Ripe queen cell being packed in thermos flask for transport to out-apiary.

in Fig. 25, half-full of foam rubber pieces to absorb about half a pint of warm water (98°F), with a thick disc of foam rubber above to hold the cells vertically in holes cut for the purpose.

It is suggested that *beginners* should rear queens from swarm cells, as just described. *Improvers* might well use emergency cells built naturally in the first instance, or from introduced combs, strips of comb with eggs, or perhaps punched cells. *Experienced* beekeepers will enjoy mastering more advanced techniques involving the grafting of very young larvae into prepared cell cups.

For improvers and experienced beekeepers the first requirement is a prosperous cell-building unit well stocked with young workers without too many other mouths to feed. Bee literature describes many ways of stocking such a colony, but in my experience there is no easier or better way of achieving this than the first stage of the simple two-queen system described on page 64. This provides a brood box of young bees, with thousands more emerging over the next three weeks, warmed by the artificial swarm underneath and at a most convenient height for handling. There will also be two frames (at each side of the brood nest) with at least one face fairly solid with pollen and honey.

Improvers

When the top box is checked after four or five days, brush the bees off very gently (with a goose feather) and mark with a drawing pin those frames containing open queen cells. It is unlikely that any sealed cells will be found, but if there are any then cut them out, as arising from larvae older than 18 hours when first given royal treatment by the workers. Assuming that only open queen cells are left, then no young queens can emerge for at least another seven days, and sealed cells can be cut out and used as required after developing for another three days. If the cells have to be transported to another apiary, it is better to wait six days in order that the young queen may be fully formed and less vulnerable to damage in transit.

The next step up the ladder of experience is to raise cells from the queen of one's choice, rather than from any stock strong enough to be two-queened. The easiest way to do this is to use a shallow frame fitted with a half-sheet of unwired foundation, and make zig-zag cuts in the lower edge. This frame should be slipped into the middle of the brood nest of your chosen breeder queen at the same time that the rearing colony is manipulated. Five days later, the first crop of cells in the rearing colony should be cut out, and this must be done most thoroughly, with bees shaken off every comb in turn. The comb from the breeder queen (now containing eggs and very young larvae) should be taken out, shaken free of bees and put into the centre of the top box of the rearing colony, which now contains thousands more very young bees which have emerged over the last five days. It also has only a small area of open worker brood to feed and can give really royal treatment to the chosen eggs/day-old larvae on the introduced comb.

The shallow comb may be left in the rearing colony until the queen cells are ripe for transfer, say ten days later, when cell tips are 'bronzed', i.e. stripped of wax by worker bees to enable the queens to bite their way out more easily. The next stage, however, is to rear more than one batch of cells. To do this, remove the shallow frame of drawn cells after three days, brush off the bees and give the frame to any top super full of bees, on another hive. Although a queen-right colony will not usually start queen cells, the bees above the queen-excluder will happily continue to feed such cells already started.

We are now at Day 8, and the main rearer

stock has no open brood at all to feed; it has a large population of young bees but also a rapidly growing population of flying bees, which will not be receiving any queen substance and may lose heart after twice having queen cells removed from them. At this stage it is recommended that the rearing colony in the top box be moved a few feet away on to a hive floor, so that its flying bees reinforce the bees in the lower box, which should at the same time be given another super; this increases the chance of a honey crop and removes the older bees from the rearing box, where the very young bees will happily draw out another batch of queen cells if given eggs or young larvae. A frame of open brood alongside the frame of queen cups would encourage them.

The planned rearing of more than one crop of queen cells makes full economic use of the young bees, and also covers the possibility of a spell of cold weather coinciding with the time of mating of the first crop. After this the bees should be left two good queen cells from which to raise and select their own queen.

Experienced beekeepers

Experienced keepers can use the basic methods already described, but with a number of refinements, such as grafting bars of queen cups with larvae from a breeder queen housed in a nucleus hive, or a small compartment of a normal hive, to slow down her egg-laying and

26 Plastic queen cells grafted with 18-hour-old larvae using sable no. 4 artist's brush.

prolong her useful lifespan. Mated queens produced in excess of demand may be stored for many weeks in the 'queen bank' described on page 69.

All over the world the practice seems to be to use a frame fitted with two bars (shown in Fig. 26), each bar carrying ten to twelve queen cups. Most operators prepare their own queen cups, but others use and re-use cups made of plastic or wood. Some 'prime' the cups with a little diluted royal jelly, but the larger operators like Norman Rice (Beaudesert in Queensland) and Ian Berry (Arataki Honey Company, Hastings, New Zealand) prefer dry grafting.

Grafting tools vary from a ground-down crochet hook (Ian Berry) to the sharpened quill of a duck feather (Norman Rice). Ron Stretford (Nelson, New Zealand) uses a Winsor and Newton artist's sable paint brush no. 4, moistened by mouth. Fred Richards (Norfolk) uses a short piece of barbed wire with one strand hammered and shaped like a miniature spoon at the end. Others use a matchstick chewed into a shaped point at one end, and it is also possible to buy a specially made tool (see Fig. 27).

27 A larva-grafting tool.

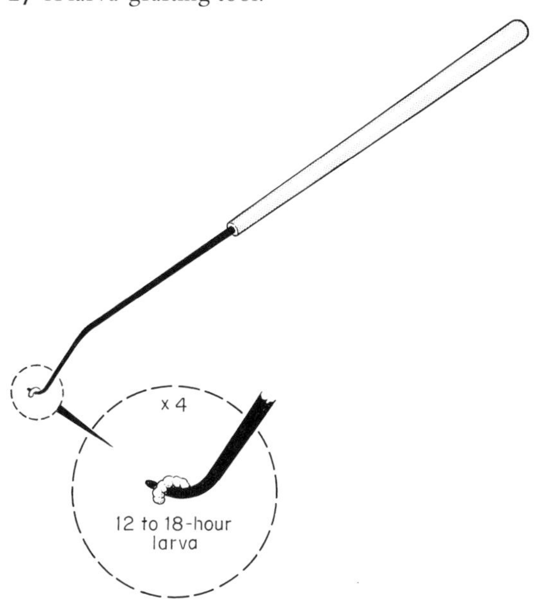

For relatively small-scale work, the wax cups may be made one at a time by dipping a wetted wooden former into melted beeswax three times and twisting off the cup between the fingers. Before doing this for the first time, cut out one or two natural queen cups, which every hive builds in May and June whether they are going to swarm or not, and shape the end of a wooden dowel-rod about $\frac{1}{4}$ in diameter or slightly larger, until it just fits snugly into a natural queen cup, using very fine glass-paper to get a smooth surface.

On a larger scale, wax queen cups may be produced a dozen at a time on a multiple former, which looks something like a wooden rake. This is first soaked in water for a few minutes, shaken and then dipped three times in melted beeswax in fairly rapid succession, each time to a slightly shallower depth. After the third dipping the cells (still on the rake), having a drop of hot liquid wax hanging from them, are gently pressed on to the wooden cell bar, itself previously painted with melted wax. The rake is lifted clear, and a quick brush-load of melted wax painted along the base of each side of the new cups on the bar; this ensures that the cups are firmly fixed to a strip of wax and can ultimately be removed, using a sharp knife to slice off a strip of wax carrying one or more mature cells.

The beeswax should not be much above its natural melting point, and can be contained in a tin or aluminium foil tray standing in a hot water bath. Norman Rice has for many years used an old electric frying pan with a thermostat, and a wide paint brush for applying a layer of wax to the bars. A normal wooden brood frame, modified with slots to take two cell bars, is used as a carrier, and it is best to put this into the rearing hive for a few hours (overnight, for example) so that the bees can 'get a sniff' of the cups and work on them to correct any unnatural feature. The frame is taken out just before grafting.

If one works very quickly the actual grafting can be done out of doors beside the hive (for up to a dozen cells, anyway), but drying out is the great danger, and most operators prefer to be sitting comfortably with a good light over their shoulder, in a warm room with a boiling kettle providing a humid atmosphere, or a greenhouse with water sprayed on the floor. If working from a normal brood comb, larvae of the correct age will be found on a comb of eggs with some open brood. If one is working on a larger scale, the breeder queen is confined to the middle third of a brood box by two vertical queen-excluders and a new frame of drawn comb given four days earlier. Most professionals agree that the ideal age for grafting is when the larva is between 12 and 18 hours old and about twice the size of a comma in this book, curled up on the cell base in a small pool of brood food. The technique of grafting has to be practised and learnt, so that hand and eye are co-ordinated, with wrist steadied by resting the elbow on the table. The larva is lifted off by slipping the tool gently under, and deposited in the dry queen cup by touching the cell base and sliding the tool back and up, the exact reverse of the picking-up motion. Even the best of beekeepers have to practise this technique a few times before it comes naturally, and it is most helpful to work alongside an experienced operator.

After grafting a frame of cups, cover with a cloth and replace in the gap from which the same frame was taken a short time before; young bees will have clustered in the space, so lower gently, and the young bees will reform their cluster on the grafted cells.

Direct use of queen cells

It is possible to re-queen any stocks with about 75% success simply by pushing a ripe queen cell between the top bars of the brood chamber, without having to find and kill the old queen. The odds are improved by protecting the queen cell as already described, and it is suggested that this be done to all stocks every two years, or earlier if one has a 'slow queen'. In this situation the cell is accepted as a natural supersedure cell and there is no check in egg-laying; if the cell is not accepted then it is probably because the bees are very satisfied with their queen anyway. There is obviously no control over drones in this case, except that a frame of drone comb can be put in the brood nest of the best colony in any apiary to be given queen cells, a month before this is to be done.

Mating nuclei

As already suggested, beginners should be content to rear a queen in the nucleus she is to head. Improvers could make up mating nuclei from two combs of bees (including brood and food), taken from any prosperous colony in May. This can take some of the steam out of a very forward stock and reduce the likelihood of an unwanted swarm. Once the queen is mated and laying she can be removed and replaced with another queen cell, after having been given a day or two to lay enough eggs to provide reinforcement for the workers. In fact a second queen cell can be given ten days after the first, so long as it is *completely protected*, for example by being enclosed in a hair-curler cage blocked at both ends, when bees will feed the emerged virgin but cannot molest her. Then, as the mated queen is removed, one end of the hair curler can be opened up and temporarily blocked with a small blob of queen candy (or a wisp of tissue paper), to delay her exit until the bees want her (a couple of hours at most).

On a slightly larger scale, it is convenient to have half a dozen mini-nucs to set up every year at the end of April, and full details are

28 Used queen cell. Queen has emerged from this cell pushed in between frame bars of mating nucleus.

given at the end of this section. Larger operators will need to use scores or hundreds of nuclei, and there are many ways of doing this. Usually a standard brood box is sub-divided to provide up to four separate compartments (as at Buckfast Abbey) or a specially constructed, longer brood box half the normal width with three compartments and half-size frames (Beaudesert, Queensland). Arataki (New Zealand) have just started using small polystyrene mating boxes, each stocked with just a large cupful of bees at the beginning of the season.

The huge Miel Carlotta enterprise in Mexico uses small wooden individual mating boxes, maintaining many hundreds in each mating apiary, and taking great care to paint vividly coloured distinctive patterns on each roof to help queens to identify their own home. Queens mate more rapidly from small units, but these can starve very quickly in bad weather and should always be supplied with a stock of candy (baker's fondant is good and easily procured), rather than fed with liquid syrup which can invite robbing.

In any serious queen-rearing enterprise it is necessary to provide at least two full-strength drone-rearing colonies, and preferably three or four, to flood the area with drones of selected parentage. Isolation is best achieved in an offshore island as the Germans do in Heligoland, the Dutch in one of the Friesian Islands, the Italians in Elba, and so on. Limited isolation can be achieved on the mainland, e.g. in a remote part of Dartmoor. The full details would be more appropriate to a specialized book on queen-rearing than a general book such as this.

Finally, a word on bee health. All the big

29 Brother Adam at his queen-mating apiary, Hexworthy, Dartmoor. (Courtesy of *The Times*.)

queen-rearers feed Fumidil B to queen-rearing stocks, mating nuclei and escort bees. Nosema can flare up when bees are stressed, as they can be in queen-rearing. Working on a small scale, I have not personally noted the need for this.

A MINI-NUC. SYSTEM

Most queen-rearers use some system involving combs (and numbers of bees) much smaller than those in a normal nucleus box. My own system uses six mini-nucs, each holding three frames, each one of which is exactly half of a standard national brood frame. During winter these 18 half-frames clipped together form nine normal deep frames and are housed with two frames of food in a single National brood box. From the end of April to mid-September the six mini-nucs are on the wall at home, used as mating colonies for a succession of queen cells. Assuming that the nucs are already stocked with bees and up on the wall, the routine goes like this:

30 Plan view of mini-nuc.

31 Mini-nuc. frames being separated.

32 Mini-nuc. and frames.

i Arrange for a succession of queen cells to be available at intervals, by any one of several different methods (covered in the section on queen-rearing). Cells from good swarming stocks can sometimes fill a gap in the planned succession.

ii As each young queen mates, allow her to lay on for a day or two to keep up the brood nest, then take her out for sale, or immediate use, or storage for a few weeks in a 'queen bank', like that in Fig. 37.

iii Replace the queen with a ripe queen cell gently pushed down between the frames.

iv Check that the little colony has enough food, and at all times keep a jar of 'set' honey or candy in the food compartment behind the combs.

v If a nuc. has too much food, take out a comb and give it to another in need, or store it for use later on. In any case replace with a fresh half-frame of foundation. Three days ago I had to take out honey-clogged combs from two mini-nucs; to-day (12 June) both replacement frames were found to be fully drawn out and laid up solid, so the young queens were taken out and used to re-queen two full stocks.

vi In a period of bad weather a mini-nuc. can reach starvation point very quickly. A half-frame of food is the best answer, but a jar of food on the rear shelf usually keeps them going. Some years ago I used to squirt syrup directly into the combs, but this can start up robbing.

vii By mid-July wasps can be a menace, and it is surprising how soon they will spot 'soft targets'. I usually cut a cork to fit in the entrance to give room for just one bee at a

33 Six mini-nucs on garden wall.

time to enter, so that 'Horatius can defend the bridge' against the barbarian wasps.

viii During the first half of September an empty National hive is placed on boxes so as to be central to the six mini-nucs, and by now only one of these is allowed to retain its queen. The minis are taken down in quick succession, frames clipped together as they are and put down one after another into the new brood box, with one normal frame of food at each end. At this time of year there has never been any argument or fighting; they set up fanners at the entrance, and with the minis no longer on the wall, the flying bees rapidly re-orient to the new hive, which is fed for a couple of weeks and then taken away to an out-apiary for the winter.

ix Next year, about mid-April, the same mini-nucs are restocked in two visits to the out-apiary, with a week in between, taking care to leave the queen on site, of course. Each mini will then raise queen cells for itself; these should be rubbed out before the first planned queen cells are introduced a day or two later. Obviously enough bees must be shaken into each mini-nuc. to look after its original brood; they will not return to their parent hive as it is too far away.

This cycle of events has taken place every year for the last fifteen, and is as much a part of spring as planting out the tomatoes in the greenhouse. It regularly produces queens for use when needed, with a few spares for friends who ring up in September and say they have a queenless stock and please can I help. The whole process is quite fascinating and adds immensely to the pleasure of beekeeping at home.

If one forgets to put in a replacement queen cell, a strong mini-nuc. will raise two or three cells for itself, and these have (to my surprise) provided some very good queens. Perhaps the ratio of bees to queen cells is the key factor, and for $1\frac{1}{2}$ frames of bees adequately to provision two queen cells is no more a problem than for 11 frames of bees to provision fifteen or more, as a swarming stock will often do.

An interesting method of speeding up queen production is to put the next ripe queen cell into a gauze or haircurler cage sealed at both ends, and gently push this down between the frames a week after introducing the last queen cell. On emerging, the new virgin will be fed in her cage, and once the mated queen is removed the virgin can be released a few hours later. If one end of the cage has a candy plug protected by adhesive tape, this can be peeled off at the time that the mated queen is removed.

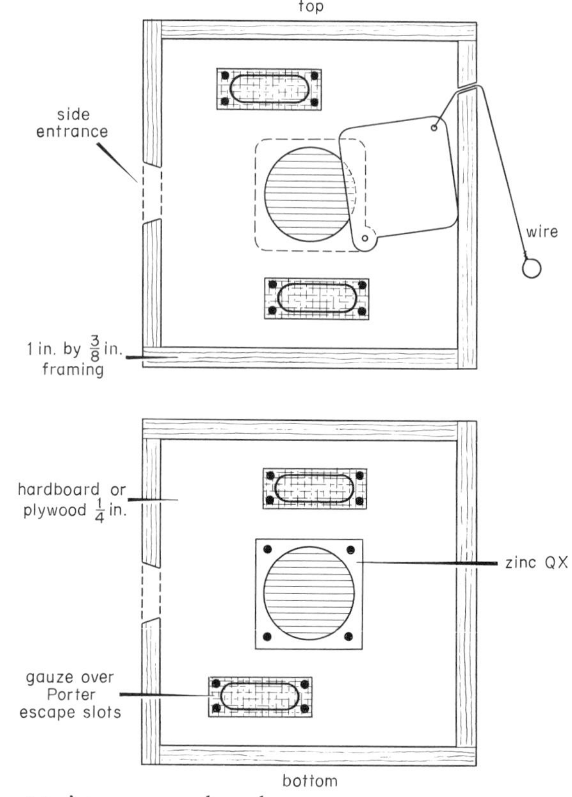

34 A two-queen board.

A SIMPLE TWO-QUEEN SYSTEM (NOT FOR BEGINNERS)

Over the years there have been many attempts in several countries to obtain a very large force of workers by having more than one queen laying in one hive. In nature two queens will usually fight if in direct contact, and only one is left. The method outlined here is a simple one, based on an old method of swarm control by making an artificial swarm in spring. The only additional equipment needed is a special two-queen board (see Fig. 34), which can be made at home by modifying an existing crown board.

No special preparation is necessary for a colony on a double brood chamber, or on a single brood chamber with a super over a queen-excluder: just feed about half a gallon of thick syrup in the second half of March, and again three weeks later. If the starting point is an existing colony on one and a half brood chambers, then it would be best to slip in a queen-excluder between the two boxes on a mild day towards the end of March, first smoking the queen down should there be any brood in the shallow box. If this shallow box has more than six frames solid with over-wintered food, take out the excess and replace with empty drawn combs centrally placed, before feeding; the object is to have a reservoir of liquid food at a time when bees often need to forage for water on cold days, and some fail to return.

By peak dandelion time (around the third or fourth week of April in Devon) any over-wintered stock in a good condition should have at least seven frames of brood and eggs in a bottom box full of bees, with some bees also clustering on at least two or three super frames. Most beekeepers attempting this system will probably have at least three or four hives in their apiary, and the strongest should be up to the required strength even if the others are not. It is assumed that the bees are in National or Langstroth hives; this method can be used with double-walled hives but obviously some provision has to be made for an upper entrance, either by wedging up one of the outer lifts or drilling a neat one-inch hole in one of them.

The operation

Move the hive three feet to one side and place a clean floor board with an empty brood chamber on the old site. Take off and cover the super, remove the queen-excluder and look for queen (see page 46 for section on queen-finding). Place the comb on which the queen is found, plus bees, in the centre of the empty brood box, together with a comb of pollen and honey on one side and an empty or partially empty comb on the other. Fill up the box with frames of drawn comb if available, or foundation if not, and replace queen-excluder and super. On top of this goes the two-queen board with central hole closed and upper entrance pointing to one side, then the original brood box, with combs pushed together centrally and two or three frames of drawn comb or foundation added at the side to replace the combs already put below. Depending on the food stocks noted, and the weather, a feeder may be put on top, but this is not usually necessary. A sheet of heat-insulating material (polystyrene or glass wool) over the top crown board is desirable for the next two weeks.

Foraging bees will return to the lower entrance, and the bottom box will house a stock resembling a normal hived swarm, except that it has a frame of brood to encourage it and the queen will continue laying immediately, so that there is no check whatsoever on the output of eggs. Should the weather be bad, these bees have the super as a food reservoir. The upper box will retain the very young bees, with more emerging every day, and will benefit from the warmth of the swarm below. Under these conditions queen cells will be raised from eggs or very young larvae. A check should be made five days later, and frames having queen cells marked with a drawing pin. It is unlikely that any will be sealed, but if they are then they should be destroyed, as arising from larvae too old to give good queens. If queen cells had already been started naturally by the bees before the operation, these should be left.

Follow-up

If the intention is to try out this system with just one hive, then nothing else need be done at this stage, except to check the upper box three or four weeks later to see if the new queen is laying. In practice it is usually obvious that the queen has mated from the behaviour of the bees at the top entrance; they fly with a new sense of urgency and take in many loads of pollen. As soon as the new queen is laying, pull the wire to open the queen-excluder panel in the two-queen board. From now on, worker bees will circulate freely through all parts of the hive, and with two laying queens the population will grow rapidly.

Options now available

a *Improvers* wishing to increase their stocks could let the two queens run on until the onset of the main honey flow, and then remove the upper box to another stand in the same apiary. Flying bees from the upper box would reinforce the main stock and maximize honey production, while the

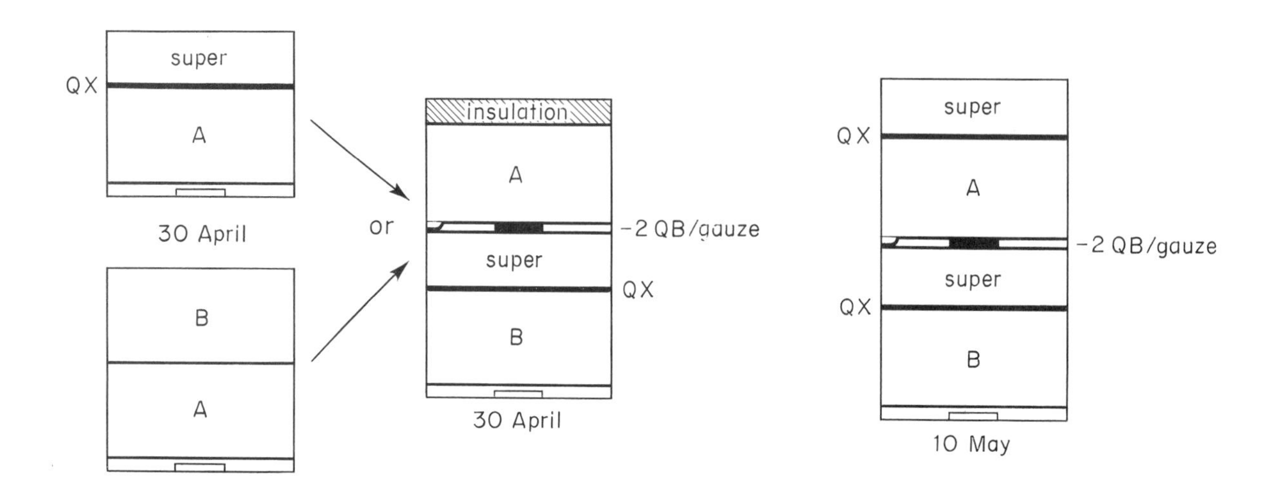

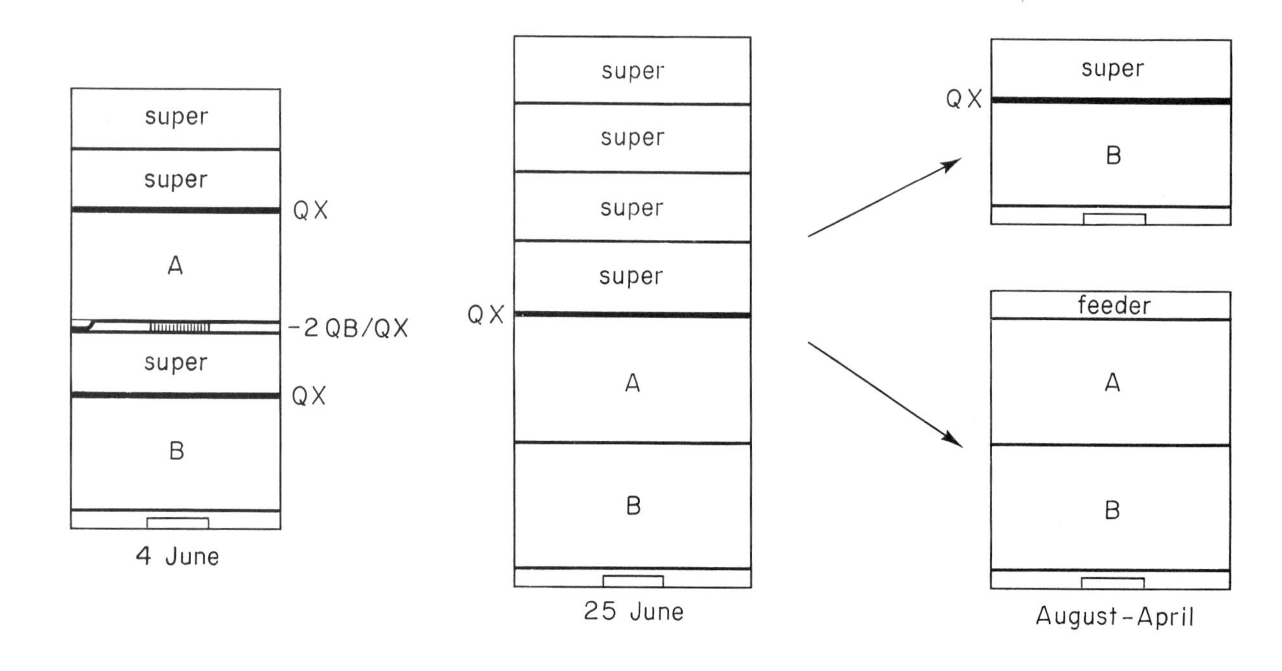

35 A simple two-queen system.

moved stock would have time to build up and store some food for themselves, and could be fed syrup later on for the winter. Alternatively, a four-frame nucleus plus ripe queen cell might be made from the upper box in May, for sale or to build up as a new stock.

b *Experienced* beekeepers could make two nuclei for sale from the top box, leaving just three frames including one with queen cell or cells and adding eight frames of foundation to replace combs withdrawn. Alternatively, as a refinement, they could destroy all queen cells in the top box after a week and put in a comb of eggs from a breeder queen, or a frame holding a couple of bars of grafted queen cells; this has the added advantage that after a week several thousand more young worker bees will have emerged in the top box.

c If no stock increase or queen-raising is contemplated, then the two brood boxes should be united before the end of June. This can be done directly as the bees are really all one stock. Just lift off the upper brood box and lower super, remove the two-queen board and put the two brood boxes together. Over them place the queen-excluder and supers, adding more as necessary. Although one might expect the two queens to meet and fight, this usually does not happen, at least not at once, and both mother and daughter may well continue laying for some time. However, by the autumn only the daughter will be there, and the colony will go into winter with a young queen.

Other possibilities

a There are advantages in working this system in units of three or four hives, feeding each in March and April (with an interval of three weeks) and choosing the strongest colony to use first with the two-queen board. Then the remaining hives can be similarly manipulated nine or ten days later and ripe queen cells from the first hive, introduced into their top boxes a few hours later. This gives the less strong stocks over a week to catch up, and has the advantage of raising queens from the strong stock of one's choice.

b Beekeepers wishing to go to the heather will find this system excellent. At the end of July the two brood boxes can be reduced to one, with a young queen and up to eleven frames of brood, plus a large population of young bees to be cleared down into the heather super.

c Many beekeepers may wish to winter on double brood boxes, but in the writer's opinion it is better to reduce to a single box early in September, give an extracted super over a queen-excluder and feed for winter. Any frames with pollen should go into the single box. This gives an opportunity to take out any poor combs and to sterilize with acetic acid those retained, ready for use next year.

Advantages noted

This method provides automatic annual requeening, or if applied every other year, ensures that no queens are retained longer than two years.

If a stock is found with natural queen cells, then swarm control is effected by this method: in the upper box just one open queen cell should be left. If the stock is very strong a nucleus could also be made from the top box at the same time.

If a colony has actually swarmed, then the swarm should be hived in a new box on the old site, with queen-excluder and super over, then a two-queen board and finally the swarmed stock. To prevent a cast, it would be best to take out all queen cells except one, preferably open. If eggs or young larvae are also present, from which further queen cells might be raised, check again after five or six days and destroy any surplus queen cells.

In two-queen stocks, the bees always seem to be in good heart; any dead bees are air-lifted and dropped well away from the hive, the bees work with energy and determination and have a general air of well-being. This is probably due to the presence of two queens, each producing queen substance at a time when some normal hives may be under stress, with elderly queens possibly beginning to fail. The absence of stress caused by such factors as shortage of queen pheromone, difficulty in covering the brood when a 'cold snap' comes along, especially when a swarm has left, and over-frequent inspection by beekeepers intent on preventing swarming, seems to diminish the incidence of nosema, chalk brood and other troubles associated with stress conditions. Queen-rearing is also carried out with less stress at the top of a hive, aided by natural warmth rising from below.

Dealing with snags

a If the queen cannot be found, then place a comb of young brood and eggs plus bees in the new box and assume that the queen is on it; proceed as usual, but next day go through the top box again. As flying bees will have gone to the lower box it is much easier to find the queen, and then simply run her into the front entrance. If she is left in the top box all is not lost, and a new queen will be raised in the bottom box. However, some of the other options are then no longer open.

b If a queen is lost on her mating flight, or for any other reason the upper box appears to be queenless, with no eggs after four weeks from the date of the operation, look very carefully for the queen, and if she is not seen, give a ripe queen cell at the first opportunity. If this cell is torn down at the side there is probably a queen there, but if no eggs are seen on the next inspection, it is probable that a stale virgin is present, i.e. a young queen prevented by bad weather from mating. Mating most commonly takes place between the seventh and twelfth days of age, but if not achieved by about the twenty-fifth day, it is not usually possible for a queen to be impregnated satisfactorily. Unfortunately in an English summer it is possible, even in June, to have a period of two or three weeks without a single afternoon being warm enough (about 64°F or warmer) for a mating flight. A stale virgin is small, active and spiteful, and resents the presence of a more fortunate sister. She is difficult to find but unless found and destroyed will attack any queen or queen cell given. Should she not be found, one answer is to give three protected queen cells placed between different frames, so that the odds are in favour of one young virgin surviving.

If a large number of queen cells is needed in the first instance, then the two holes in the screen board should have gauze panels on both sides, to reduce contact between bees on either side. Using gauze on one side only can sometimes result in just one or two queen cells being started.

BANKING QUEENS (NOT FOR BEGINNERS)

Some years the best weather comes early in summer, and a glorious April and May is followed by three to six weeks of rain, high winds and temperatures too low for queens to mate, even at midday or early afternoon. I well remember one summer a few years ago when, after a fine May, the daily maximum air temperature in June failed to reach even 60°F for 25 consecutive days, and not one of a batch of a dozen virgins mated. There are also other years (like 1983, for example) when a cold, wet, early summer has suddenly changed and given way in mid-June to a prolonged hot spell with temperatures well into the seventies day after day for weeks. For reasons like this it is sometimes said that our climate is unsuitable for queen-raising, and that it is best to import from America, Australia and New Zealand. However, the current mood in beekeeping circles is strongly against importing queens, so what is one to do?

The answer is to start the first batch of queens towards the end of April, and then at intervals during the summer through to the end of July, if necessary. Even in the worst summers, there are some warm days. In practice, queens are needed for routine re-queening at the end of the season, also as occasion may demand at any time, so that some method of keeping young mated queens until needed has to be used. On a small scale this is usually done by using pairs of nucleus hives with three to five frames and uniting two whenever a queen is taken from one of them. However, this results in an increased number of stocks which may not be wanted, and also locks up many frames of bees in hives not likely to produce a surplus of honey.

About 20 years ago I started using a queen bank made by cutting horizontal slots in a frame of drawn comb and fitting in hair-curler queen cages, each holding one mated queen and no workers (see Fig. 36). Centrally placed in a deep super over a queen-excluder in a

36 Queen bank, home-made.

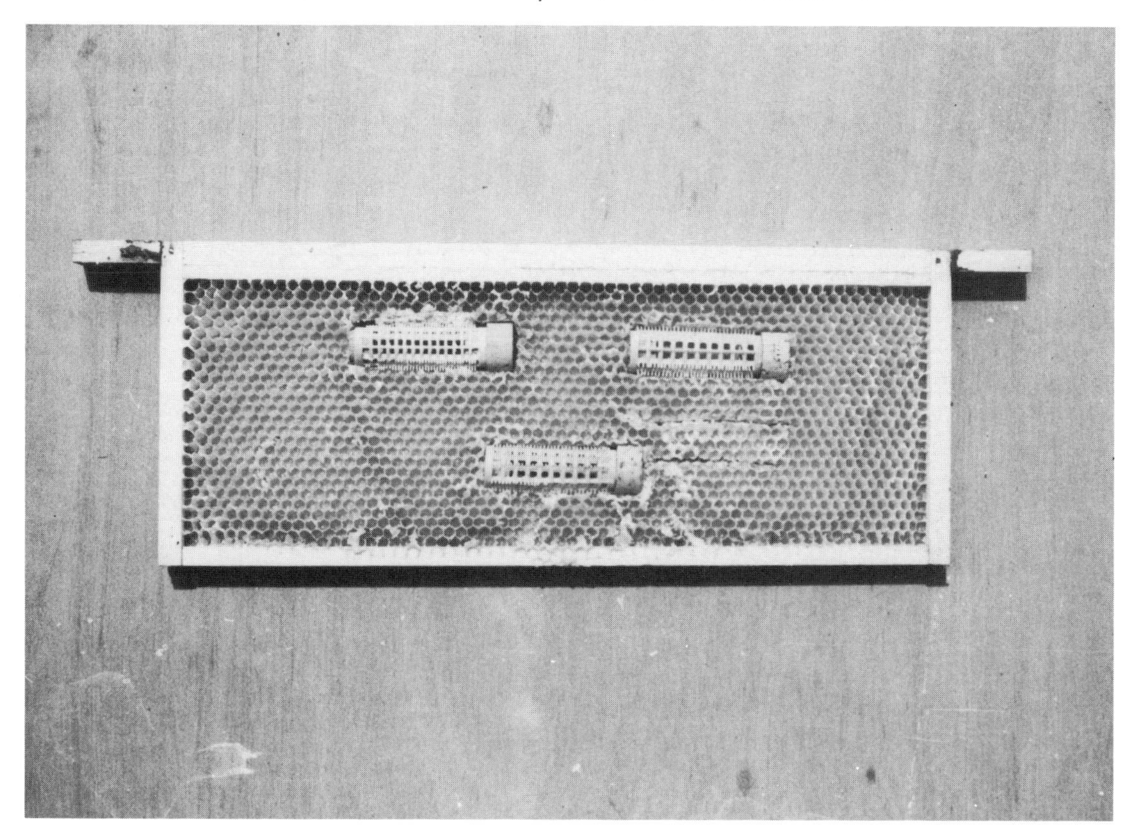

37 Queen bank: frame holding 24 queens in super of queen-right colony. (Courtesy of Norman Rice, Queensland.)

38 Method of stuffing queen cages with escorting workers before putting in queens. (Courtesy of Norman Rice, Queensland.)

queen-right colony, queens were held successfully for many weeks, though never beyond mid-October. A shallow frame, in a top super, can also be used. Working with Mr Norman Rice in Queensland in February 1984, I found a much neater and simpler version of this system, used for banking and holding many queens ready for quick filling of large orders. As shown in Fig. 37, it consists of deep frames modified by two sets of parallel, opposing strips of wood near the base and mid-frame, to take two rows each of 12 wooden queen cages (as used for dispatch through the post), held vertically. Two such frames, centrally placed in a deep super of normal frames, over a queen-excluder on a queen-right colony with plenty of young bees, can hold 48 young mated queens for weeks at a time. In practice, young queens are usually allowed to lay for three to four days in their mating nucs, then caught and run into cages, without any escorting workers or candy. The normal corks blocking exits should be covered with slips of adhesive tape (such as Sellotape) to make sure that workers will not be able to release the queens, and a little queen candy smeared over one of the gauze panels of each cage to attract workers in the first instance. The wooden sides of normal transit cages provide sufficient barriers to prevent contact between adjacent queens, and a drawing pin marker on the top bar is used to indicate frames containing queens, which will be fed by workers. When queens have to be sent through the post, cages can easily be taken out and the queens transferred into fresh cages supplied with candy at one end, ten to twelve workers run in as escorts to each queen and both ends (cork/candy) covered with adhesive tape for transit. I am indebted to Mr Rice for showing me his technique.

Four

Early Summer – swarming

A swarm of bees in May is worth a load of hay;
A swarm of bees in June is worth a silver spoon;
A swarm of bees in July is not worth a fly.

WHAT IS SWARMING?

Swarming is basically the reproductive instinct of the colony, and without it bees would have vanished from the face of the earth countless centuries before primitive man evolved. In nature, catastrophes such as forest fires, cold wet summers, falling trees, and attacks by bears, woodpeckers, etc., probably eliminate one wild colony in five every year, and without swarms to maintain the number of colonies bees would have become extinct. Within each colony the reproductive instinct keeps up the number of bees, but there also has to be the instinct to swarm in order to keep up the numbers of colonies.

In practice, swarming means the building of several queen cells in May, June or July and the subsequent departure of the old laying queen with about half the bees, to find a new home and start a new colony from scratch. From the queen cells young virgin queens hatch, one of which will survive to mate and lay in the old home. One or more smaller swarms (called 'casts'), headed by virgin queens, may also leave before things settle down again. In any case the chances of a crop of honey from that hive have been lost (or very much reduced), so beekeepers view the procedure with disfavour and dream of 'non-swarming stocks' and of 'preventing swarming'. Although some stocks swarm less than others, such dreams are not realistic answers to the situation, and a better approach is to minimize swarming by good basic management, and then to 'go with the bees' to satisfy their basic instincts without actually losing a swarm.

Swarming factors

The actual swarming impulse can be triggered off by a number of factors, such as unsatisfactory performance by an old queen, who amongst other things may fail to produce enough queen substance to satisfy a very large work force. Research work in the 1950s showed that whereas at least one hive in four could be expected to swarm in an average year throughout the country, this figure was reduced to one in twenty for stocks headed by young queens (one year old or less).

Overcrowding can also be a main cause, either on account of too small a hive or because of brood nest congestion in even a large hive. This is especially likely to be the case if a queen has reached an early peak of laying (perhaps up to 2,000 eggs a day), and then has to reduce her laying drastically because the brood nest is choked with nectar because supers have been put on too late. This early peak of egg-laying produces a glut of young nurse bees a month later, at a time when a frustrated queen has been unable to find room to lay more than perhaps 300 to 500 eggs a day. Thus there is a huge force of young nurse bees with too little work to do: this can lead to swarming. The weather pattern also has an effect, and when a cold, wet fortnight in June comes after a warm, sunny spring, frustration and overcrowding tend to induce swarming, which then takes place on the first dry, warm day.

Good management can greatly reduce swarming, but never completely eliminates it. In any case there are always swarms from unknown sources that have to be rescued from lamp posts in busy streets, rose bushes in frightened householders' gardens and so on. Sometimes swarms from elsewhere even come uninvited to the home apiary and ask to be taken care of. Some swarms can even read, or so it seemed three years ago when one came from a distance and occupied an empty hive of mine with the word 'swarm' chalked on the side, meant to be used for that purpose should a swarm be caught! The swarming season is usually reckoned to run from mid-May to about the first week in July, but sometimes continues into August. They usually come out between 11.30 a.m. and 3 p.m. on a warm day, especially following several days of cold or rain.

39 A nice swarm, just taken from a tree.

Types of swarm

Up to mid-June the probability is that it will be a prime swarm, that is, a swarm headed by a mature queen that brought the stock through the winter and built it up in the spring. Such a swarm usually pitches low down and often within a few feet of the parent stock, although after a few hours or up to a couple of days it will move on. Later in the summer there will be more after-swarms or casts, usually smaller and settling higher up and further away, headed as they are by more active young virgin queens. Then there are mating swarms, often characterized by not forming a compact cluster and sometimes even sprawling over three or four square feet of roof or ground. Lastly, there are panic swarms, when shortage of food, flooding, invasion by ants, etc., has driven a stock to abscond as a last resort; such a swarm may even be queenless. To put the matter in perspective, out of every 100 swarms that one comes across, roughly 50 will be prime, 40 casts and seven or eight mating swarms; the balance will be absconding, often queenless swarms.

Management for reducing swarming

Obviously a management system can be evolved to take account of the main factors, and many beekeepers do quite well by re-queening every year and accepting small swarm losses. These losses can be still further reduced by putting on the first super before the end of April, to give room for the bees as well as to decrease the risk of a flush of early nectar choking the brood nest. Also an April inspection may show that combs of unused pollen and honey are blocking the expansion of the brood nest, and exchanging these for empty combs (perhaps from a stock needing more food) is good practice. Quite apart from congestion, it also helps if young bees can be given work to do, such as making wax and building new comb. This can be achieved by removing two or three old brood combs each year and putting in frames of foundation. If a double brood box is used, this is best done in the upper box.

The technique of brood-box reversal is being widely used in America now, and can be employed whether the bees are in a double brood box or on 'one and a half'. Quite simply, this involves putting the upper box down on the floor and bringing up the bottom box every 10 to 12 days from about the second week in May.

40 Swarm on a wall.

Early summer – SWARMING

Queen cups

Almost every strong stock of bees will make a few queen cups towards the end of April or early May, about the time that the first batch of drone cells are sealed over. This is not a signal of intention to swarm, any more than the building of drone cells is, but rather a reasonable precaution to enable swarming preparations to be made as quickly as possible should they feel the need at any time. Queen cups should not be cut out but carefully noted, as they provide a useful indication to the beekeeper. For example if one of the central combs has two or three empty queen cups, it is extremely unlikely that the stock will swarm for at least nine days; at any subsequent inspection look first at the same frame and if the cups are still empty, look no further. If, on the other hand, there is an egg, or a larva floating on a bed of creamy royal jelly, one can work out when the cell is likely to be sealed and know how many days are left in which to make an artificial swarm, or employ whatever swarm control technique is favoured. If the queen has a clipped wing, then no swarm can leave until the first virgin has emerged, and again one can calculate the number of days left.

We do not know exactly what prompts the queen to lay in a cup, but one theory is that the workers hem her in as she passes close by. Although a queen cup is larger than a worker cell, in fact slightly larger in diameter than a drone cell, a queen will still lay a fertilized egg in it; this is possibly because queen cups are usually built in among worker brood rather than drone brood. Sometimes workers will remove eggs from queen cups if the mood of the colony is against swarming. Queen cups look rather like small acorn cups, hanging downwards.

SWARM CONTROL

However good your management system, inevitably some stocks will build queen cells and prepare to swarm, and some control is necessary to avoid loss. The actual cutting out of queen cells every nine days is not recommended, for a number of reasons. Firstly, the morale of the colony suffers and less work is done while more queen cells are built. Secondly, it is easy to overlook just one queen cell tucked into a corner and covered with bees, and then all your work achieves nothing. Thirdly, the process may have to be repeated several times and you will end up with a frustrated colony. A better approach is to go with the bees and exploit their swarming urge to the mutual advantage of bees and beekeeper. There are many way of doing this, and it would take too much space to describe them all. In general terms, some method of making an artificial swarm is recommended, such as those described elsewhere, e.g. using a two-queen board, Taranov's method (see page 85) or the orthodox method, as quoted on page 83.

There is also the alternative of operating, say, 25 hives with queens not more than two years old, and accepting that four or five of them will swarm, instead of having 20 hives and spending hours of work on swarm control to get the same yield of honey. One has to balance capital costs against labour. In a suburban garden, of course, there can be no such choice – one must avoid the bees swarming on to a neighbour's property.

Another method is by isolating the queen and allowing the bees to re-queen themselves. This can be done when queen cells are spotted by making up a small nucleus with the frame that the queen is on, plus a frame of food. Put this nucleus box some distance away and give a very small entrance, after shaking in a frame of young bees. Go through the parent stock carefully, destroying all queen cells except one open one with a larva. Push brood frames together and add two frames of foundation, one at each side of the brood nest. Mark the top of this frame with a drawing pin placed on the frame top directly above the queen cell. Five days later check that it has been sealed and destroy any other queen cells that may have been made since. The bees can now satisfy their reproductive urge and rear a new queen, while the old queen will be on hand in case anything goes wrong. If this be done after mid-June there will be no loss of honey, as eggs laid later than this produce bees maturing too late for the main honey flow anyway.

Swarm-catching – equipment needed

For strength, lightness and general convenience, there is nothing to beat a good, old-fashioned straw skep, but failing this, a stout cardboard box of about $1\frac{1}{2}$ cu. ft (with no loose flaps) will serve. Also have an old hessian sack, a piece of hardboard about 2 ft square, a goose feather or pigeon's wing to use as a brush, a small bow saw, pair of secateurs, and an extending ladder on the car roof-rack. Smoker, veil and gloves should always be taken, even if seldom used. An old brood comb (empty) is useful to lure bees out of very awkward spots and a queen cage (hair curler and cork) in case the queen is spotted. (See Fig. 41.)

41 Swarm-catching gear: 2 skeps, screenboard, hessian sack, white sheet, bow saw, secateurs, smoker, old brood comb, queen cage, goose wing, rope.

Shaken swarms

My ideal swarm is one weighing about 5 lb (about 20,000 bees) hanging in a compact cluster a bit larger than a rugby football from a gooseberry bush about 3 ft off the ground. Here the technique is as follows: *a* spread the sack on the ground over the hardboard more or less under the cluster; *b* hold the skep or box under and close to the swarm; *c* give the branch a sharp shake to dislodge the bees so that they fall in a heap into the skep or box; *d* quickly invert the skep on the sack, placing a pebble or piece of stick under one edge to give an entrance. Some bees will fly and go back to the branch but the great mass of them, hopefully including the queen, will be inside or at least on the sack. Within half a minute bees outside will start running in and soon a dozen or more will be seen at the entrance and up the side of the box fanning, with heads down and tails up, Nassanov gland exposed, to broadcast the 'come in' message. When the fanners are hard at work, shake the branch again to dislodge the knot of bees still there, and as they fly most of them will pick up the pheromone (scent conveying a message) and home in on the fanners. Now is the time to accept the cup of tea offered by the grateful householder, and come out twenty minutes later to find 98% of the bees inside the skep. Fold the corners of the sack up and over the skep, pick it up and put it in the car boot. On a very hot day, it is

42 Swarm being hived in a five-frame nucleus, half of a double nuc.

advisable to tie a thin rope round the skep over the hessian and invert to allow ventilation via the coarse, open weave of the material. Before leaving, explain to the householder that there may be a fistful of bees on the branch next day, left behind; this will save the nuisance of a phone call next day saying that there is another swarm in the same place. These few bees usually disperse within 48 hours. As a short cut, there is no reason why the hanging swarm should not be shaken directly into the hive it is destined to occupy, if this is convenient, as it might be in this particular example.

Brushed swarms and driven swarms

Obviously a brick wall, telegraph pole or tree trunk cannot be shaken, but the technique is much the same. Hold the skep close to the trunk under the swarm and gently brush the bees downwards in three or four steady strokes, so that they fall into the skep, then invert on the sack at the foot of the tree as before.

Occasionally, a swarm will settle in a thick bush or closely trimmed hedge, from which it can neither be shaken nor brushed. Here the technique is to trim off some twigs with the secateurs so that the skep may be placed on the hedge or bush, over the swarm and as close to it as possible. Now puff smoke gently under the cluster and shake or tap the branch very gently if possible. The bees will retreat upwards: as soon as a few reach the skep they will start fanning and draw up the rest, especially if the skep has been used a few times before and has traces of wax in it from previous occupation. This same technique may also be used for bees on a wall, roof or thick bough of a tree, whenever a skep or swarm box can conveniently be placed above and close to the cluster.

Special cases

There are many variations on the theme, but perhaps one or two may serve to illustrate the general technique. Bees in chimneys are notoriously difficult, but if they have arrived in the last 48 hours they can often be dislodged by smoke. If the householder is advised to burn damp brown paper or rags in the fireplace, giving much smoke and little heat, they will often cluster somewhere in the garden where they can easily be taken. If they have been in the chimney a week or more, they will probably have a brood nest and will not leave. At this stage I usually chicken out and explain that I am a beekeeper, not a steeplejack!

A swarm 20 to 30 ft (6–9 m) up a tree and at the end of a bough can be awkward, but if the bough is not too thick, proceed as follows: *a* trim off any small branches beneath the branch holding the swarm, then gently saw through the branch close to the trunk until it starts to give but is held by about one-eighth of its thickness; *b* grasp the bough 3 ft or so out from the trunk and pull steadily with a rigid arm, as you slowly come down the ladder. With a modicum of luck the bough will come down as if hinged at the saw cut, so that the swarm is then accessible.

The swarmed stock

During a spell of fine weather the prime swarm will probably have come out up to a day after the first queen cell has been sealed over, which means that the first virgin queen will emerge about six or seven days later.

At this stage there are at least three possibilities:

a The first virgin out will hunt for and destroy other virgins in their cells, aided by worker bees who will tear down queen cells from *the side*, so that just one queen is left to mate and take over.

b Two or more virgins may emerge about the same time, or before the first queen can locate and organize the destruction of rivals. In this case they may fight until only one is left to lead the colony. More often the bees themselves chase the young queens around and choose one, subsequently handling the others roughly so that they hide in corners, sometimes squeeze through the queen-excluder or even come out on the alighting board in order to escape.

c The bees may protect one or more queen cells by clustering tightly on them. When the young queen attempts to bite her way

43 Queen cells torn down at side after first queen or queens have emerged from other cells.

out in the usual way, by making a circular cut at the lower end of the cell, they will wax up the cut, leaving a small slot through which the entombed queen may extend her tongue to be fed. Only when the first virgin has flown out with the second swarm (called a cast) will the next virgin be released, either to kill those remaining, or to fly out a week or more later with a second cast. By this time the swarmed stock has been so weakened that there is certainly no hope of a honey crop, indeed it will have to work hard to build up enough bees and food to survive the winter.

What should a good beekeeper do about all this? The main consideration is that no further swarms should issue, and that as soon as possible the swarmed stock should resume normal business. Yet if all queen cells but one are destroyed at once the bees may, for at least four or five days, construct more and then throw a cast. Hence normal practice is to wait for five days and then go through the swarmed stock thoroughly and destroy all cells but one, leaving only *one open queen cell* with a fat larva seen to be there on a deep bed of royal jelly. The reason for this is that an apparently good sealed queen cell may in fact be empty, as the 'trapdoor' of an emerged cell may have been pushed back and then sealed with wax by a zealous young worker. Sometimes such a cell may be found with a dead worker inside, suggesting malice but more probably an accident. Again, the sealed queen cell may contain an imperfectly developed virgin, perhaps detached from her bed of royal jelly by rough handling of combs. After five days there cannot be any eggs, or any larvae younger than two days around which another queen cell might be made. The alternative to waiting five days is to choose a good open queen cell immediately, marking the frame top with a drawing pin, and destroying all others at the same time. Then five days later re-check and destroy any made subsequently.

It usually takes much longer than one expects before the young queen in a fairly strong stock is mated and laying, and there can be a period of three weeks or more before eggs are seen. If a young mated queen is available, then she can be introduced about five days after the swarm has emerged, after destroying all queen cells, and the break in egg-laying reduced to about a week.

General points

Some mention should perhaps be made of the situation when bad weather delays the issue of a swarm. Sometimes when a swarm is delayed five days or more, up to three or four virgins may emerge in the excitement, in time to fly with the swarm. A check through the swarmed stock will then reveal perhaps a number of naturally emerged queen cells, and there is no easy way of knowing if a virgin is running around free. In this case perhaps the best course of action is to destroy all queen cells but one open one with a very young larva, and go through the stock again five days later to destroy any others. Should a virgin be seen, then all queen cells should be destroyed.

If the original queen had her wings clipped, then no swarm can leave until the first virgin flies. A swarm may come out with the old queen but will return when she fails to fly. When a prime swarm leaves a hive there are usually several combs of sealed brood, with 1200–1500 young bees hatching every day. After seven to nine days this means that there are another 10,000 bees in the swarmed hive, easily enough to allow a second swarm, or cast, to go. During a spell of reasonable June weather I have known wild colonies (in a hollow tree in one case, a village church steeple in another) to throw a prime swarm on one day, a cast nine days later and a second cast nine days after this, all three alighting on exactly the same site (a wooden fence in the first case, a 15 ft holly tree by a grave in the other). If one knows the hive which swarmed and captures the swarm, then the method of hiving on the old site described later on is most effective. Otherwise the treatment just given should be followed.

How does one get to know about swarms? The standard method is to notify the local police in April of your phone number and that you are a beekeeper willing and able to come out at short notice to collect a swarm that may be causing a nuisance. It pays also to leave a similar message at the local council office. You are then likely to receive emergency calls about swarms on street lampposts, on the front wheel of a parked car, on a vegetable stall in front of a shop, in a chimney, on a wall, on the canopy of a petrol filling-station and so on. You will also get swarms hanging peacefully from rose bushes in quiet gardens, as well as false alarms. Whenever possible it helps to speak directly to the person who rang the police. Ask when the swarm arrived, whether it has clustered (looking like a large coconut or rugby football) or is still flying. Some of the calls will then be established as wasp nests, bumble-bee nests or just a large number of bees foraging on a cotoneaster bush, for example, and the householder's mind can be set at rest, or advice given on what to do about the problem.

Calls in April are often about a few dozen harmless solitary bees emerging from their tunnels in a sandy bank, or from a tile-hung side of a house. Calls in July are often about wasp nests, and may be referred to the local council pest-control officer.

Bait hives

From May to July it really does pay to have a spare brood box containing some old combs, perhaps one or two black and gnarled with drone cells that are no longer good enough for a working colony yet very attractive to bees. Such a box of combs (plus makeshift floor and roof) in a corner of your garden, or better still in a friend's garden 300 yards away from your apiary, will very likely catch a swarm that might otherwise have been lost, and the bees can easily be shaken off comb by comb into a nucleus or empty hive of your choice.

Experience shows that swarming bees prefer to choose a new home away from their own site, so that the best place for a bait hive is not in the actual apiary itself. A bait hive at home may entice a swarm coming from one of your own hives, but is rather more likely to be occupied by one from a distance. What are the ethics of this? In law, bees are *ferae naturae* or wild creatures, and so another beekeeper has no title to a swarm he believes to have come from his hive. On the other hand, an accepted code of conduct among beekeepers is that swarms should be returned if one is certain of the owner. Placing a bait hive 100 yards from the home apiary of Buckfast Abbey would be regarded as 'sharp practice', but a swarm that may or may not have originated from a known apiary a mile away can fairly be ascribed to a hollow tree in a nearby park. There is a strange fascination in setting up a bait hive, something like planting a lobster pot in the sea off a rocky coast.

MAKING THE MOST OF A SWARM

The modified Pagden method

If it is known from which hive the swarm came, then that hive can be moved three or four feet to one side and turned 90°, while the swarm is hived on foundation or drawn combs in a new box placed on the old site, with excluder and supers on top. Flying bees from the old hive will reinforce the working force and the chances of a honey crop are good. After seven or eight days, or as soon as a considerable number of young bees are seen to be flying from the old hive, first turn it back 90° and move it slightly closer to the swarm, and the next day move it to the other side of the swarm and again turn it so that the entrance faces away. This will shoot another 5,000–12,000 workers into the swarm, thus increasing the chances of a honey crop and decreasing the likelihood of the old stock throwing a cast (second swarm).

A variation which works well is to place the old brood box over a screen board with side entrance on top of the supers on the swarm, thus saving the need for another floor, roof and stand. This also gives access to the old frames at a convenient height and makes it easier to check the queen cells and make a nucleus from the swarmed stock if so desired, while the warmth coming up from the swarm below reduces the risk of chilling brood. Later on, when the new young queen is laying well above, the box can readily be united to the swarm to re-queen it.

If there are two or more hives in the garden it may not be obvious which has swarmed. A convenient trick is to capture six to ten bees from the swarm in an empty matchbox, drop in half a teaspoon of flour and shake vigorously for a few seconds. Then open the box and watch the hive entrances to see which one the 'white bees' are entering. The flour does no harm; it will be combed off as pollen would, and the shaking merely disorients the bees temporarily so they forget they have swarmed.

Package bees

Have ready a deep box of drawn combs, placed over a framed wire excluder on a floor; hive the swarm through the entrance in the usual way and leave overnight. In the morning the box, containing 98% of the workers, may be lifted off gently and placed on a second floor, while underneath the excluder will be found the queen with a small knot of bees, including drones. The main body of queenless worker bees may then be united through a sheet of newspaper to a working colony to boost their workforce, used as package bees, or given a ripe queen cell to re-queen them. In the meantime, the old queen may be given an extra cupful of bees to keep her going on a couple of frames in a nucleus box as an insurance against future needs.

Food frames

A swarm can also be used to produce a surplus of stores on brood frames. As these will contain pollen and honey as well as ripened sugar syrup, they will be invaluable as outer food frames in established colonies being made ready for winter. Also, four- or five-frame nuclei may be found in October to have enough bees but insufficient stores, and the addition of two heavy food combs on the flanks will see them comfortably through the winter. The point in this technique is to feed heavily and continuously, unless there is a big nectar flow, as the wax builders have to draw out the combs and once they stop secreting wax (due to a check in the income of nectar or syrup) they will not readily resume.

Repair work

A newly hived swarm or cast has a wonderful aptitude for repairing old combs or drawing new ones, and some old frames may have the drone comb scraped away down to the mid-rib (with an old worn kitchen spoon) or even cut right away, and a swarm well fed with syrup will be happy to put on a patch of new worker cells. For this work, a cast (headed by a virgin queen) is preferable to a prime swarm. A prime swarm, if headed by an old queen, may draw out a patch of drone cells.

Booster techniques
A swarm in a skep or upturned box will establish itself on combs entirely of its own making as readily now as 100 years ago, especially if well fed through a small hole in the top. Place the skep or box colony alongside a producing colony, and when some four or five weeks later it begins to gain in strength, move it to the other side of the hive, thus shooting several thousand workers into the producing stock. The process may be repeated 14 days later if the flow is still on, by moving the skep right away into a quiet corner and feeding it to build up for the autumn, when the bees may be driven up into a nucleus box and fed for the winter. The ancient technique of bee driving is described on page 120.

Stocks for next year
The old saying that 'a swarm of bees in July isn't worth a fly' may be true as far as prospects of honey that year are concerned, but as a July swarm is most likely to be a cast headed by a young virgin, it is worth hiving and feeding up to give a stock headed by a young queen for next year. There is also time to observe the brood pattern and assess the merit of the new queen, and replace her before late autumn if she is not seen to be doing well enough.

New colony
Of course, an early swarm may also be hived in the usual way in a box of frames fitted with foundation, to grow as quickly as possible into a working stock. The tendency to abscond will be greatly reduced if one frame of drawn comb can be included. The books speak of using a frame of young brood from another hive, but this may not always be practicable or desirable and an empty old brood comb will hold the bees just as strongly. The swarm should be fed sugar syrup (3 lb sugar to 2 pts water) while they are drawing foundation and until they have a reserve of food at the tops of the brood combs, but do not feed for the first two days. Prime swarms often supersede within a few weeks; in any case they are usually headed by older queens, and it is good practice to re-queen before the winter, if they have not done so themselves.

Disease from swarms
The official doctrine is that a stray swarm may carry disease, and if hived in one's apiary might infect other hives. It would be unwise to deny the possibility, but my experience of over 30 years with never less than 10–15 swarms (not my own!) hived per year without any disease arising, suggests that the risk is small. Undesirable characteristics (such as bad temper) may well be present, in which case the swarm should be re-queened before winter. In practice disease is more likely to be carried by old combs than by swarms, for several reasons. A colony weakened by disease is less likely to be strong enough to throw a swarm, and also any infected honey taken with them is likely to be used up rather than stored, especially if one waits 48 hours before feeding the swarm. The one common disease most likely to be carried by a swarm is nosema, and it is well worth while to add Fumidil B to the sugar syrup given to a swarm after hiving as a routine precaution.

Swarming behaviour
During a period of three days in the middle of May one year, four swarms arrived in my garden, but definitely not from my own hives. One managed to get into a pile of supers awaiting transport to an out-apiary; another into an empty hive with some drawn comb and some foundation, actually waiting for a swarm should one be found; a third pitched on a blackcurrant bush and the last one on the woodwork of a garden frame. I had carried out

a swarm check on all hives at home a few days before but assumed that I must have missed some queen cells. On a further and most painstaking check I found two hives making swarming preparations but no larvae more than one and a half or two days old in any queen cell. The other eight hives were clear, some with empty queen cups and one or two with no emerged drone brood even. All were crowded with bees, several needed extra supers and four had one super full with a second well on the way. The queens in all four swarms were unmarked; all my queens are marked, and were seen to be so in April. Fortunately all the swarms were in my own garden and not in a neighbour's, for who would have believed me if I said they were not mine? This type of behaviour has happened here before, in fact one Sunday morning a few years ago I was standing with a friend among my hives and we heard the roar of a swarm approaching from a distance. It hovered overhead and then settled on a low branch of the old Bramley in the garden, about 15 yards from the nearest hive.

Do swarms pick up the flight path of thousands of foragers and fly in with them to an apiary? Or do the scout bees perhaps do this? I don't know of any hive within three-quarters of a mile of my home apiary, but the bee traffic radiating from here is considerable, especially in the warm weather. My own private theory is that bees gossip and mine had spread the news about a kind-hearted beekeeper who feeds his bees candy in February if they are short, sugar syrup in March and puts on the first super of drawn comb in April.

Late swarms

A phone call from Mr Len Brimacombe of Brixham reported the taking of a swarm on 18 September; it weighed about 2 lb and was clustered at eye-level on a cotoneaster bush. Eye witnesses stated that it had arrived around midday. Next day we received a phone call about a 'bees' nest' in a hedge at Teignmouth and set out sceptically expecting to find wasps, but in fact it was a genuine cluster of honeybees which had built a mass of external comb in a hawthorn hedge about 7 ft from the ground. By cutting around with secateurs and then carefully sawing through two supporting branches, I hoped that the nest could be removed intact, but in the event the wax was not strong enough to support the remaining piece of branch and the nest had to be moved in two parts. Once home, the bees were shaken on to drawn brood combs in an empty hive and the wild combs laid above the crown board; there were no eggs or brood, but the queen was seen. Within two or three days the bees had taken down their own honey and were sealing it over; at this stage they were fed thick sugar syrup in addition. The central areas of the middle three combs had obviously been bred in more than once, so presumably the swarm had stayed in the hedge most of the summer. Impressed by the unusually even, white cappings which these dark bees produced, and by the way they worked early and late, I took great care of them, and the following summer they produced some of the finest comb honey I have ever seen.

On 25 September a phone message from the local council reported a swarm of bees flying in large numbers at an address in Garden Road, Ellacombe, Torquay; the swarm was said by local residents to be a danger to children at a nearby junior school. On inspection the 'swarm' proved to be intense foraging activity on several square yards of Michaelmas-daisies flourishing in an otherwise derelict garden, in full sun with an air temperature of about 70°F. The facts of insect life and how the birds and the bees work were explained to the satisfaction of three very worried ladies.

General points

Mention has been made more than once of the use of old brood combs: in practice many beekeepers accumulate a number of such combs when replacing the usual two or three combs a year with foundation, so that it is assumed that they came from healthy stocks. Further to this, it is good practice to sterilize such combs with acetic acid before re-use.

It must always be remembered that a swarm only has the food it carries with it, and if it encounters two or three weeks of cold, wet weather (quite possible in an English summer), then it must be fed generously with sugar syrup until it becomes established with reserves of food.

Finally, a word on the sheer pleasure a swarm can give: in the challenge of taking it and overcoming new difficulties in doing so; in the thrill of shaking the bees on a sheet or board and seeing them marching in, with fanners quickly at work, Nassanov glands exposed to call in the flying bees, and the signs of good housekeeping even in the first few hours as old pollen, broken pieces of comb, etc., are brushed out in an orgy of spring cleaning; in the fascination of opening up after a few days and spotting the first patch of eggs and open brood as the nursery develops, and the pure white or light gold of the newly drawn comb.

If you have the time, and a clustered swarm is conveniently placed, get a chair, call for a tray of tea and settle down for a fascinating hour watching the swarm from a range of 12 inches. Apart from dances by scout bees, you will almost certainly see the queen, as she usually walks in and around most of the time, appearing from a fissure in the cluster, walking on the outside for a few seconds and then diving in again. If you have an empty match box or a hair-curler cage, or even an empty half-pound honey jar, it is not difficult to scoop her up, plus a few bees. Once you have the queen, she can be caged on a comb in a hive where the swarm is required to go and there will be no difficulty about hiving the swarm successfully.

A newly emerged swarm will have about 3,700 bees per pound, but a swarm not taken for several days will have less honey and perhaps 4,000 to 4,500 bees per pound. An average swarm, of about 3½ pounds, will have between 12,000 and 15,000 bees.

MAKING AN ARTIFICIAL SWARM – BEGINNERS IN THEIR SECOND YEAR

By the beginning of June over the whole country some stocks will have swarmed already, and many more will be making plans to do so. It would be must unusual, however, for the beginners' four-frame nucleus of last year, headed by a queen barely a year old, to have done so yet, although there are probably drones flying around at midday, and perhaps one or two queen cups on central frames. The major task this month is to increase to two stocks by making an artificial swarm, and there could not be a better time of the year than the first week in June. Although it is easier and more pleasant to do this on a sunny day in warm sunshine, it can be done in the evening, or in fine rain if necessary. Do not delay beyond the first week on account of weather, and be ready with a second hive consisting of floor, brood box containing eleven frames fitted with foundation, crown board and roof.

Move old hive

First smoke the hive in the usual way and then slide it sideways along the stand, replacing it with the new hive, into which some of the flying bees will soon start to go; take out the central three frames of foundation and lean them against the side of the hive. Now take the roof off the original hive and place it upside-down on the stand between the two hives, ready to receive the super. Insert the hive tool between super and queen-excluder at a corner and gently lever up; puff smoke into the gap and then run the hive tool along the base of one side of the super to separate it from the excluder, and lever up again at the next corner. Lift the super slowly with a gentle, turning motion and place it diagonally on the upturned roof, so minimizing the risk of crushing bees. If there is any tendency for the excluder to lift with the super, use the hive tool at the other corners as well. Now take the excluder off, peeling it away like a large piece of plaster if it is an unframed zinc one, otherwise treating it like a crown board; in either case check that the queen is not among the bees on the under-surface before propping it up against the entrance of the hive.

Queen and two combs to new box

You now have to find the queen, so be prepared to take some time over this if necessary. Put the inspection cloth on and puff some more smoke into the entrance, left, right and centre: this should ensure that the queen is off the floor and most likely to be on the combs. Roll back the cloth to expose a couple of frames, take out the end one, as you have done before, and lean it against the side of the hive. It is unlikely that there will be either eggs or brood on this frame, but in any case have a good look for the queen before putting it down. Please do not shirk this exercise of hunting for the queen. This is the *pons asinorum* of beekeeping; see that you cross it and you will be a beekeeper! When you see her, carry the comb she is on, plus bees, and place it gently in the central gap of the new hive, keeping her in view as you move across, with her side of the comb uppermost. Now look for the comb containing sealed brood plus bees and put this next to the queen's comb. Check that there are no queen cells on either comb; if there are, cut them out with the corner of the hive tool. Now push up the frames of foundation, fitting the spare one in at the side. Return to the old hive and push up the frames towards the middle to close the gap, keeping the brood nest as an integral whole before replacing the end frame and the remaining two frames of foundation at the ends. The old hive can now be given the crown board and roof that came with the new one. On the new hive place the queen-excluder, super(s) plus crown board and finally the old roof.

Swarming problem solved

We now have a situation in which most of the flying bees of all ages will rejoin the queen in the new hive, forming what is virtually a swarm. On the two combs transferred there will be enough nurse bees to look after the small amount of larvae present, while the sealed brood will provide a few thousand young bees to look after the next generation and keep a reasonably balanced population in the meantime. In these circumstances there is little likelihood of any swarming problem this season, but next season the queen should be replaced, as in her third year a swarm would be likely, even probable.

New queen raised

In the old hive there will be enough young nurse bees to look after the remaining brood, and queen cells will be drawn around eggs or very young larvae in perhaps four to ten worker cells. In about 11 or 12 days the first young queen will emerge, and given reasonable weather will mate and start laying three and a half to four weeks from the date of your manipulation, in good time to build the colony up to full strength well before the autumn. Any surplus of honey is likely to be from the swarm in the new hive, where the queen will have been laying continuously. The super should have been fairly heavy at the time of the swarm and so will be a reservoir of food, but if the weather should stay bad for more than a few days it would be wise to feed, as the bees have to draw out nine frames of wax foundation.

General points

It would be an advantage to have fed the stock generously early in April, so that honey from the fruit blossom and sycamore in May is more likely to be stored in the super, or even two supers in a good year. It has been assumed that the colony had made no swarming preparations. If any queen cells were found, that is, sealed queen cells or queen cells with larvae in them (as opposed to empty queen cups), then great care must be taken to cut out any from the two combs transferred, or there is a chance that the queen might leave with a swarm. Queen cells in the old box should be left.

Work for June (Second-year beginners)

a Make an artificial swarm;
b in a honey flow give a second super; in a long cold, wet spell be prepared to feed the swarm;
c have ready a spare brood box and frames in case you hear of a swarm;
d watch the two hive entrances and note the build-up of flying bees in the old hive after a few days;
e attend apiary meetings organized by your local beekeepers' association.

TARANOV'S SWARM BOARD

The great Russian beekeeper Taranov realized over 40 years ago that for two or three weeks before swarming there is a steady increase in the number of unemployed young bees in a colony, over and above the number needed to feed larvae, cover the brood and build wax. He aimed to take out this surplus of young bees before swarming point is reached, and devised a very neat way of doing so. His method is still a good way to prevent swarming, and very useful if one has to go away for three weeks in May or June.

The actual board is very simply made, just two pieces of wood each about 20 in long and about 12 in wide, hinged or loosely nailed at one end and with a strut or wedge to hold the open end level with the alighting board. It helps to nail a rough strip of wood 10 in × ½ in × ½ in just below the open end to give something for the bees to hang on. In operation, the board is placed to give a 4-inch horizontal gap between it and the hive entrance, and a 3-ft square white cloth is placed on the ground at the hinge end, overlapping that end of the board itself.

The idea is to shake or brush almost all bees from every brood comb on to the cloth, and return the combs to the brood chamber; a few young bees may be left on the combs to look after open brood, but the younger flying bees will also do this. Also brush bees from super frames. Some bees will fly straight back, but most will walk up the board and divide themselves into two groups: the swarm bees plus the queen hanging in a cluster from the end of the board, and the bees which fly across the gap back into the hive. After about an hour the board plus cluster may be taken away, just as if it were a normal swarm, and even if hived in the same apiary the bees will remain in their new hive. The original colony will then raise a new queen itself, or may be given a ripe queen cell to cut down the queenless interval. A few days in which no eggs are laid is no bad thing at this time of the year, and gives the bees an opportunity to clean up generally. If carried out early in June, the complete gap in brood-rearing duty after nine days may coincide with the beginning of the main honey flow, when young bees will be diverted to foraging earlier than they would if they had larvae to feed, and the production of thousands of bees just too late for the main flow will be avoided. The new queen will still have plenty of time to build up a force of young bees for overwintering.

This artificial swarm will be better than a natural swarm in two ways; *a* it will have a fair number of very young bees not yet able to fly; *b* it will not have so many of the older bees which in natural swarming are attracted by the swarming process to a certain extent and join in. Thus when hived it will work for about a week longer than a natural swarm before the inevitable decrease in population becomes obvious, until the next generation of young bees is reared. A swarm like this will work energetically to establish itself, and if placed alongside the parent colony may be used to reinforce it after a couple of weeks by moving it away a few feet during the honey flow, when its foraging bees will join up with those in the main colony. As shown in Fig. 23, supers and queen-excluder are left on the original stock.

Five

High Summer – long days and busy bees

DRONES

Their importance

The very word 'drone' has passed into the English language as a symbol of laziness, of a creature which does not work but still enjoys all the good things in life, useless except for the one function of mating with a young queen. Even this, it was said until recently, needed only one drone. Some older books on beekeeping recommend drone traps, to reduce the drone population and prevent them from eating too much of the honey crop.

We have known for some years now that a young queen normally mates with six to nine drones, sometimes in quick succession on a single mating flight, but also sometimes on two or even three mating flights in the space of perhaps two days. If she should only mate with one or two drones, perhaps because of low temperatures or bad weather over the crucial period between her fifth and twenty-fifth days of age, then her spermatheca will not store the four to seven million sperms necessary for normal fertility sustained over two or three full seasons, and she may become a 'drone-layer' during her first winter. This spells disaster for the colony as workers gradually die out and no young ones emerge in the spring to replace them. Fortunately it happens much more often that a queen runs out of sperm during summer when the colony can raise a new queen by supersedure, frequently unknown to the beekeeper, unless he marks his queens.

The mating role of drones is therefore now recognized to be of paramount importance, but why should they be raised in hundreds, perhaps even thousands, when only a few are needed? Very experienced beekeepers agree that the bees seem happiest and work best when they have a normal population of drones, and that hives with such a balanced population produce most honey. Any attempt to trap and kill off drones is certainly counter-productive, as the food used to raise them is far greater than that which they need in adult life, and the colony will only be stimulated to produce still more drones. So drones help to sustain a colony's morale. They also make a valuable contribution to the warmth of the hive: indeed, the very fact that they only fly in the warmest part of the day means that for the rest of the time they are present in the hive, helping to maintain the temperature of the brood nest. Several keen observers (with glass-sided hives) have reported seeing drones herded like sheep by workers to keep unwanted draughts away from the seams of brood. Even as larvae, their metabolic heat production makes a contribution to the warmth of the nest, as of course all larvae, and resting workers, also do.

On a fine afternoon in summer any apiary will have hundreds of drones in flight, and the sweet sound of their musical 'drone' (best enjoyed by a beekeeper half-asleep in a hammock or deckchair in the same garden!) is one of the great pleasures of life. But why so many, when a few dozen should be enough? One answer probably is this, that in nature there are no apiaries and colonies of bees are much more widely distributed, so that often each stock has to produce enough drones for itself. I have often watched swifts in high summer wheeling and diving above my home apiary, obviously feeding on flying bees, and uttering high-pitched, shrill notes of appreciation; when young queens have been lost on mating flights I have mentally blamed these birds but have no proof. It may be that, among hundreds of flying drones of approximately the same size, the occasional flying queen is less likely to be noticed and picked out by birds as a large and tasty morsel. So, whether as husbands, furry draught screens, hive hot-water bottles, decoy targets or just hive morale boosters, drones are necessary. We have to consider how best to prevent them being raised in unnecessary numbers, and in awkward places in the hive, distorting our lovely combs of regular worker cells and filling in spaces we wish left clear.

Drone control

As a general rule it will be noticed that colonies headed by young queens produce fewest drones. In a queen's second year rather more drones will be seen, and in her third year more still as the likelihood of swarming increases. So here again the advantage of regular re-queening is seen, with no queen allowed to continue heading a honey-producing stock beyond her second year. On the other hand, there are no known techniques by which drone production can be completely prevented, nor would this be desirable. Even if we use perfectly drawn worker combs, or start again with worker foundation on new frames, some of the cells will be converted to drone cells, and drone comb will also be built out at the edges, and between brood boxes if double brood boxes are used.

As always, it is best to 'go along with the bees', and this can be done by giving them some drone comb at the edge of the brood nest in April, on a frame with wide spacers to allow the drone cells to be built out to their normal depth without interfering with the surface of adjacent combs. A full frame of drone comb is not necessary, and about a third to a half of a normal deep brood frame will be sufficient. This can be achieved by fitting a shallow sheet of worker foundation in the top half of an empty deep frame and leaving the bees to fill in the lower half with their own comb, which will normally be of drone cells. In subsequent years this comb can be manipulated to be the third one in from one end and left there or placed next to the edge of the brood nest, wherever this may be in April. Any beekeeper who really enjoys detailed work on frames can go one stage further and fit a strip of drone foundation below the worker foundation. Fig. 45 shows such a prepared frame with a narrow strip of drone foundation neatly wired in to form the lower third of a composite frame. The two wax sheets can be joined by running a *warm* soldering iron lightly across a slight overlap, before fixing the composite wax sheet in the frame.

The possession of one or two such drawn frames is also an advantage if one is attempting to rear queens unusually early in the year, when some drone comb introduced into the brood nest of a very forward colony in the second half of March will ensure a supply of drones for

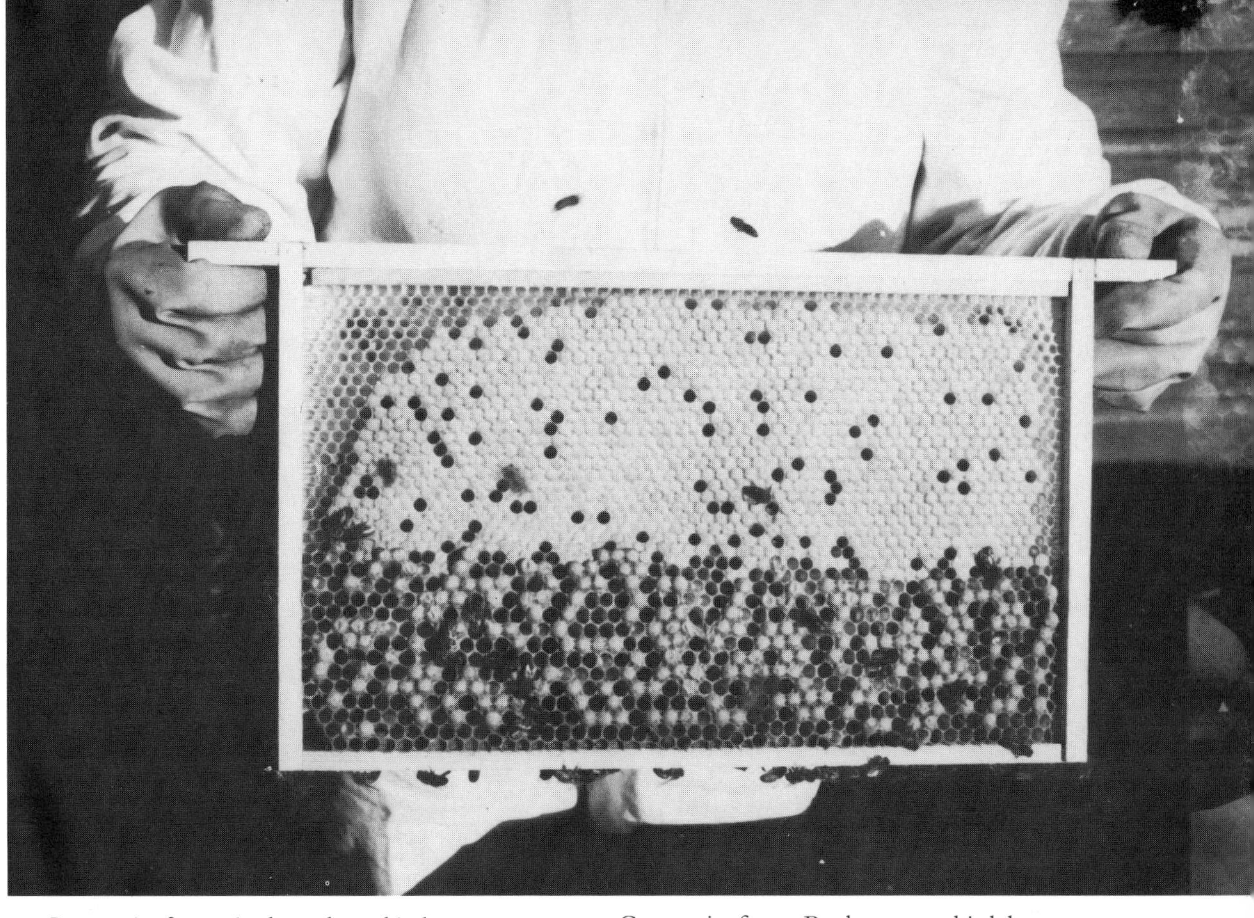

44 Composite frame A: drone brood in lower third of comb.

45 Composite frame B: about one-third drone and two-thirds worker foundation.

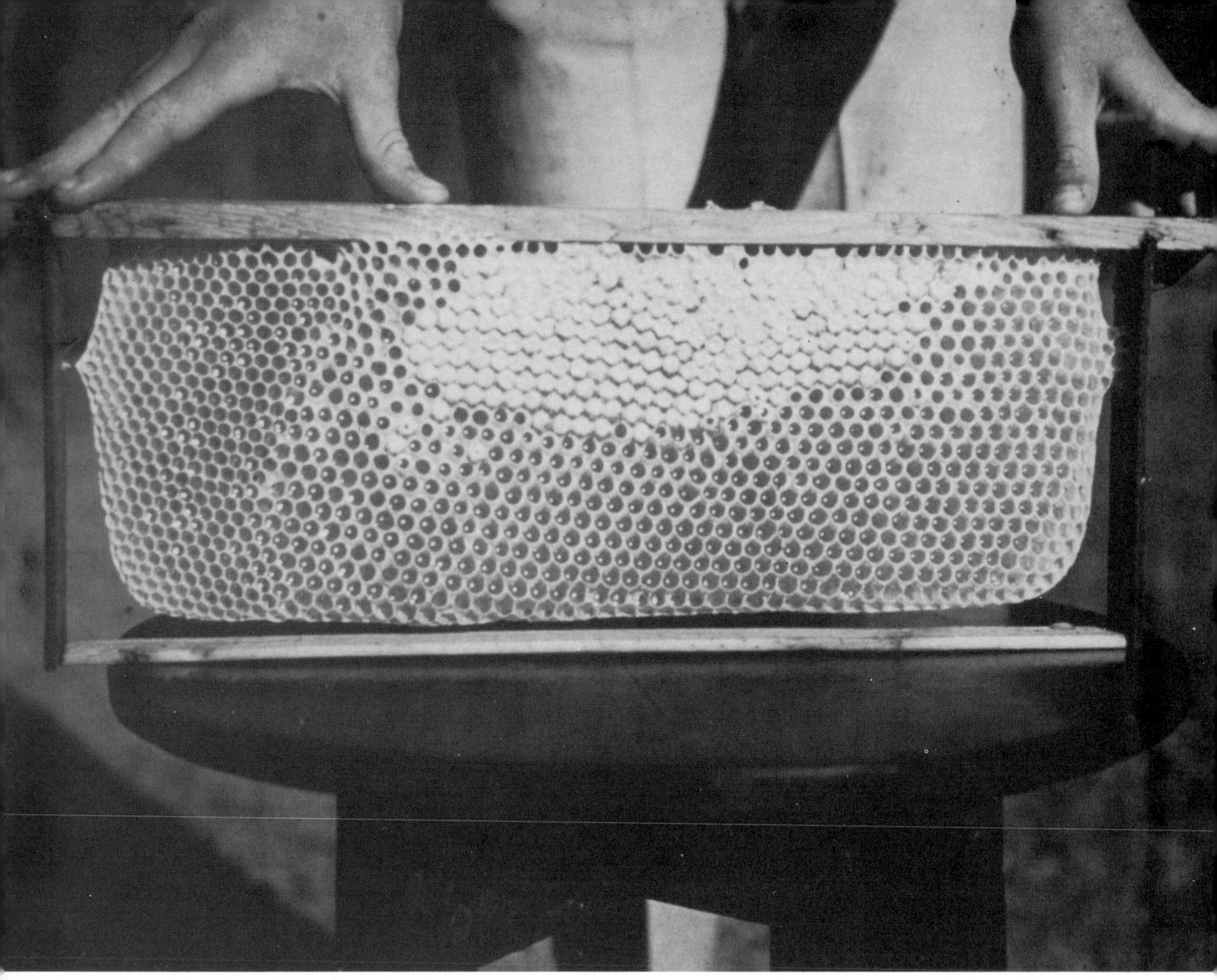

46 African comb worked by Adansonii bees, Zambia 1962.

queen-mating five weeks later, when drones might not always be available naturally. Should the dreaded varroa pest ever reach Britain, these drone combs will be valuable, as explained in the section on 'Healthy bees', page 177.

Drone comb in double brood boxes

Many beekeepers work on one and a half or double National brood boxes, and will have experienced the nuisance of drone comb constructed between the two boxes, often in the space between the frames. Sometimes frames immediately above each other are stuck together along their full length, and it is difficult to separate the boxes or even to lift out a frame from the upper box. When one does, a long row of drone cells is torn open, with white larvae exposed and often ruptured with milky white fluid spread over the top bars. Even without putting in drone comb, this difficulty can be averted by placing the upper box so that its frames are at right angles to those in the lower box. This might seem unnatural, but in practice is seems not to matter to the bees. Many beekeepers think that if there is a bee-space of a little over $\frac{1}{4}$ inch (about 7–8 mm) between two boxes, this will be kept clear. In fact bees respect a bee-space around their nest, on all sides and above and below, but any such space *within* the nest is liable to be filled up with unwanted comb. Obviously this particular technique is not possible with Langstroth or WBC hives.

WORKING FOR HONEY IN THE COMB

Balanced management

In years past, it was the practice to put section racks over crowded brood boxes and accept the disadvantage of a fairly high ratio of swarming. More recently the tendency has been to go for extracted honey only, using the same supers of drawn combs from year to year. Somewhere in between these extremes lies a sensible management scheme giving the bees opportunity to make wax and draw comb at a time when they have a natural urge to do just this. Some books say that this is wasteful and that it takes up to 20 lb of honey to produce 1 lb of wax. This is nonsense, the true figure being 6 lb, but the point is that a large population of young bees gorged with nectar will produce wax anyway, and if they have no opportunity to build comb there will be wax scales dropped on the hive floor and blown out of the entrance, odd pieces of burr comb built in unwanted places, or else they will be rather more likely to swarm and build their wax combs in a hollow tree or in someone's chimney. In any case, the weight of beeswax used in comb building is less than $\frac{3}{4}$ oz (only about 20 g) per 1 lb (454 gm) of honey stored, i.e. well under 5%.

It is good practice to get the first normal super (of drawn combs) on when dandelions are fully out (third or fourth week in April), if not already on, and as soon as the super is well occupied by bees to add a shallow box of thin, unwired foundation for comb production. New combs will be more readily drawn if immediately over the brood nest, but there is then the risk that pollen may be deposited in

47 Some good combs built naturally on starter strips of foundation, Zambia 1962.

48 New honey in an old bell jar. The bees shown in Fig. 42 went on to put 14 lb of comb honey in this bell jar, which was surrounded by glass wool insulation and a fitting empty brood box to keep warm and dark.

49 Skep of natural honeycombs.

central combs, and not all customers appreciate the health value of this, so the comb super may have to be put on top, and kept warm by a square of polystyrene or a couple of carpet squares over the crown board. The top super of drawn comb will serve as a reservoir for a flush of nectar which might otherwise congest the brood nest, and clusters of young wax-working bees will readily form under this in a comb super.

The comb super

The standard technique is to fit shallow frames with full sheets of very thin, unwired wax foundation, but an alternative is to fit only thin starter strips; this is enough to get straight combs built, so long as the hives are upright, and the young bees prefer to cluster without being divided by partitions of full-depth wax sheets. Also, natural combs are then built without a noticeable mid-rib, and connoisseurs will approve the difference. Personally, I use one $4\frac{1}{2}$-inch square of section foundation, cut into three strips hanging freely from the top bar. In Central Africa, with no access to manufactured wax foundation, I used to tack a thin strip of wood 14 inches long, of cross-section $\frac{1}{2}$ inch square under each top bar and paint it with melted beeswax as shown in Fig. 7; occasionally the bees would build curved combs running across 2 or 3 frames, but usually they built straight combs down from the bars. It was interesting to read, years later, that T. W. Woodbury of Exeter was doing this in the 1850s! It helps to have a frame of drawn comb at each end of the super, as only in a very good year will the bees draw out and fill the end combs to a saleable standard.

Using a swarm

A large swarm in mid-June, just before the main honey flow, can be an excellent producer of comb honey. To take advantage of this, have ready a shallow box of wired foundation but including in the centre just one frame of drawn worker comb, preferably of old dark wax. Over this place a queen-excluder and then a second shallow box fitted with thin foundation or starter strips, crown board and roof. Run the swarm into the entrance in the usual way and after 24 hours feed a gallon (no more) of thick syrup. This will be used up in three or four days, partly by the wax-working bees and partly to provide stores in the new brood chamber. Nectar coming in during the next month will go mainly into the comb super. Be satisfied in an average year if you get 25 lb of comb honey from the central seven or eight combs, but be ready to follow up success by putting on a second super if the swarm is very large or if there is an unusually good nectar flow. Two small swarms, taken over a period of three or four days, may be thrown together with no fear of fighting. Of course success does depend on the weather, and if there should be three weeks of strong winds and heavy rain there will be no comb honey, and the swarm will need feeding.

Section honey

Fewer sections have been produced in recent years, probably because of the wastage involved in a poor season when most are imperfectly finished; also the wooden section frames lose their fresh, clean appearance when stored over winter and put on again the following summer. Bees have a natural dislike of working in small compartments, and have to be really crowded before they will readily accept and work in an old-fashioned section crate. The best method of getting a crate of sections drawn out and filled is probably to use the swarm method just described. An alternative is to use the round Cobana sections, which sell readily as a luxury honey pack. They also have two other advantages:

a the absence of woodwork means no staining on the frames, or at least any stain is easily removed; and

b the bees seem to work more readily in the round.

After all, nothing in nature is built on the square. Figs 50–52 show how Cobana sections are fitted with foundation and worked.

50 Cobana sections. Wax foundation fitted.

51 Cobana sections filled by bees.
52 Cobana sections. Finished product ready for sale.

General points

It is normally good practice to get permanent combs drawn out with frames close-spaced, in order to avoid the irregular comb often built in empty spaces. Under the conditions required for comb honey, however, it is better to use frames spaced more widely, say ten, or even nine in a super normally taking eleven. This ensures wider, fatter combs which are necessary to get the 8 oz weight for a standard cut-comb container. A small amount of irregular comb is acceptable. Equal spacing of nine or ten frames may be achieved by driving thin panel pins through very small holes drilled in the frame lugs, by tacking brads (stud-headed nails) into the wedge sides of Hoffman frames, rapid finger-spacing by estimation as the supers are places on the hive, or by using a super with castellated frame supports.

A cut-comb super should be taken off as soon as possible, before the clean white cappings have been discoloured by innumerable bee footprints. The whitest and most attractive combs are usually produced by darker bees, which leave a small air space under the wax cappings. Bees of a lighter colour more often overfill the cells, leaving no air space and giving a slightly darker, almost greasy appearance, as the honey is in contact with the capping. There is no difference in flavour or bouquet. The 8 oz combs can be cut with a sharp knife dipped in hot water, but it is very much easier to use a Price's comb cutter, designed for the purpose. To be saleable, the top surface of the comb should be completely capped. Honey in combs not completely

53 Unwired comb being cut into 8-oz sections.

capped can be gently spun out in an extractor (low speed only with unwired combs). Partly drawn combs can also be used in supers for the heather in August. Comb honey is more trouble, but the quality of the product and the price it fetches (almost twice that of run honey) will compensate. Better results are obtained from hives headed by queens not more than two years old, as hive populations tend to be smaller with older queens.

One great enemy of comb or section honey generally is the larva of *Braula coeca*, which burrows just under the cappings and causes obvious disfigurement, though with very little real damage. Usually the pest is quite rare, but if you notice it when going through the brood chamber, some action must be taken if comb honey is to be produced. (See section on 'Healthy bees', page 175.) For run honey, the occasional small tunnel may be removed with the cappings and probably pass unnoticed.

READING A COMB

The first big advantage in practical ability comes when a beekeeper can lift a frame out of a brood chamber and assess from it, and perhaps two others, the condition of the colony in detail. When learning, it is sometimes helpful to have an audience, or at least someone to respond to what you say and prompt with questions. Here are some of the main features one should look for and comment on.

Sealed brood

A good comb of brood will probably have some sealed honey in the top corners, more in the corner further away from the entrance. There may also be some cells of pollen between the honey and the brood. The brood area itself should be compact and oval, except that a really good comb may extend over most of the rectangle and so lose its oval shape. In summer

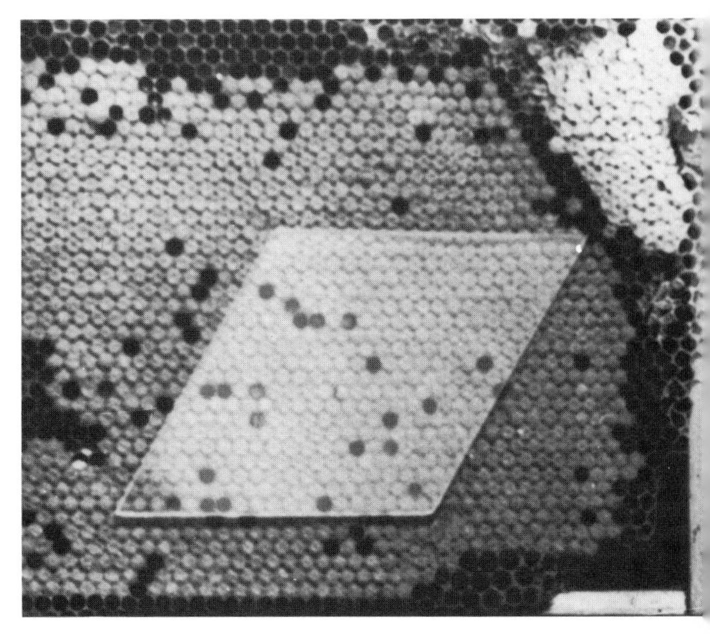

54 Brood viability meter, covering 400 cells. This brood is 95% viable.

there should be some whole combs solid with sealed brood, but in spring there are more likely to be concentric rings of open brood outside an inner oval of sealed brood, showing the development of the nest as the season advances. In any case there should be not more than a few 'pop-holes'; the actual proportion of these may be measured using the rigid plastic meter shown in Fig. 54, but the fewer the better. What is termed 'pepperpot' brood, with a large number of pop-holes (usually caused when workers recognize and remove larvae with some abnormality), is indicative of a poor queen, possibly inbred. Drone brood scattered around in individual, high-domed cells usually indicates an old or imperfectly mated queen, at the end of her useful life. Brood cappings tend to darken as they age, but cappings on older combs will always be darker than those on new, yellow combs, as wax is recycled from neighbouring cells. A comb of sealed brood with a few empty cells, possibly next to one or two young workers actually biting their way out, will most likely produce thousands of newly emerged bees in the next day or two, and would be ideal for making up a nucleus. Somewhat lighter cappings, especially next to an area of pearly-white larvae (open brood), suggest that no young bees will be emerging from that particular comb for at least a week or ten days.

Eggs and larvae

The actual age of larvae can be estimated to within a day with practice. Even the age of eggs can be estimated, and a patch of eggs still very upright on the cell base would indicate that they had been laid during the preceding 24 hours. The 3:6:12 day ratio means that with a steady rate of laying there should be twice the area of open brood and four times the area of sealed brood as that of eggs. If the proportion of cells with eggs is larger than a quarter of the sealed cells, then the queen has increased her laying rate, and vice versa.

With experience, the difference between ordinary empty cells and those recently polished in anticipation of a queen laying in them can be detected, and suggests that a new queen is just about to start laying, possibly after a spell of post-swarming queenlessness. Eggs are most easily seen when the sun, or main source of light, is behind you. I found great difficulty in this respect last year when demonstrating in Holland in a large bee house with an open front, working of necessity from behind hives on a bench with strong daylight coming into my eyes from the front.

Previous history

Sometimes a little detective work can pinpoint the day when a queen was removed or accidentally killed, possibly by careless manipulation. For example, if a colony expected to be normal turns out to have no eggs

55 Frame of honey and pollen. The end with more sealed honey will have been at the rear of the hive, away from the entrance.

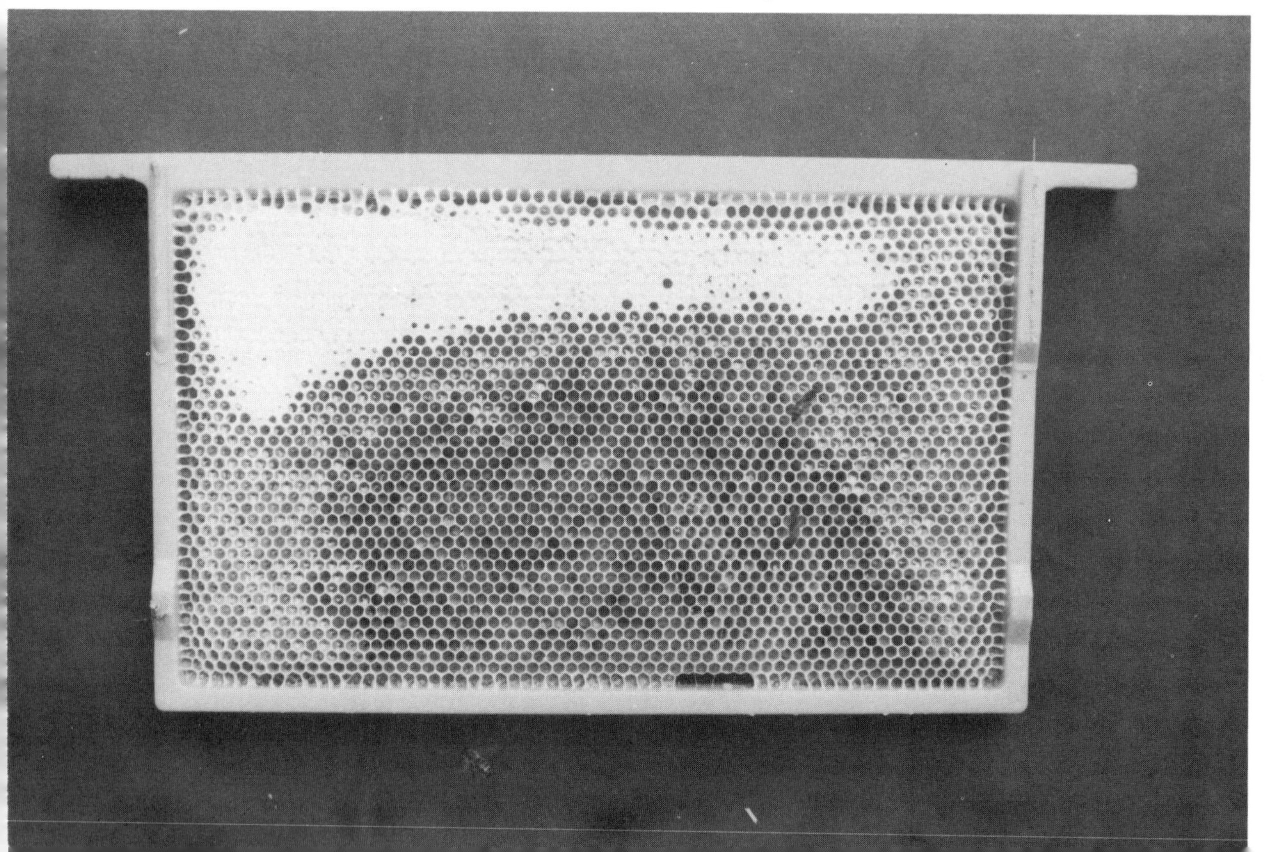

56 Good comb of sealed brood.

or open brood at all, the age of the youngest sealed brood can be estimated by opening up a few cells in different places with the corner of the hive tool and noting the age of the occupants. On one occasion I remember having to demonstrate in another county and the first hive I came to, expected to be normal, had no eggs or larvae and no queen. Checking the sealed brood here and there showed that most had purple eyes, but the youngest cell occupant was a white pupa with well-formed exterior and only a faint tinge of eye colour, so due to emerge in about seven days. On my pronouncing that the queen was lost exactly a fortnight ago, the local beekeeper said, 'That was the day Mr — opened up this hive!'

Swarming intentions

These can be read, and briefly, one is looking for queen cells and queen cups, also any sign that egg-laying has diminished, or even stopped altogether, suggesting that the bees are slimming down the queen ready for her to fly with a swarm. Much may be learned from the presence or absence of queen cups from May to July, and they should not be torn down automatically, except where this is necessary to see into them. When three or four empty queen cups have been noted on central combs, especially in conjunction with a fair-sized area of eggs, then no real intention to swarm is yet there. If in May or June there are few eggs yet plenty of sealed brood, with some drones on the combs, plus one or two cups with eggs or young larvae, then swarming is likely within days, as soon as the first queen cell is sealed over. The presence of sealed queen cells at an inspection may mean that the swarm has already left, and the absence of fresh eggs, or any eggs at all, would reinforce this, as would a relative shortage of bees in the super or supers above. When a swarm has left, many bees come down from the supers into the brood chamber, so that the brood frames may still appear to be as crowded with bees as ever. Sometimes the presence of eggs will refute a suggestion that the stock swarmed some days ago. For example, a swarm was collected by

57 Open queen cell.

my wife five days ago from an apple tree in the orchard up the Teign Valley where I have an apiary, and today when I checked all brood combs in all hives I found eggs in every hive, no sealed queen cells anywhere and the nearest approach to swarming was one stock with queen cups containing eggs. So the swarm came from the apiary of a colleague about a mile away.

Evidence of recent supersedure, in the form of a queen cell almost completely torn down, probably placed towards the centre of a comb, may be a valuable clue when re-queening is contemplated, as an indication of the possible presence of both mother and daughter queens.

Other factors

Large areas of pollen are usually found on the faces of combs at both edges of the brood nest, and this is useful when working into a colony from the outer combs, in order to check on activity, amount of brood present, and so on. In this way, working from both sides, the size of the brood nest, e.g. five combs, may be determined without actually disturbing it on a cold and windy day in mid-April.

The actual behaviour of the bees is part of full comb-reading and character assessment of a colony. Bees which run all over the comb, hang in lumps at the base (and sometimes fall off) are a nuisance. In a well-behaved colony the bees will sit quietly on the combs, and a good queen will ignore the beekeeper and go quietly about her business as usual.

The actual physical condition of the combs is also important, and note should be taken of the amount of space capable of carrying worker brood. A bad comb may have large knobbly areas of old drone comb along the lower edge and side, with large areas of worker comb broken down and changed to drone comb.

Again, here and there one may find a comb with cells too deep on one side and not deep enough on the other; this may be because the original sheet of foundation was not well fitted, so that the mid-rib is off centre over a large area. In a weak stock there may be evidence of wax moth damage, especially on the outer combs. For these and other reasons a hive apparently having eleven frames of poor worker comb may actually have less than the effective cell space of only six or seven good frames, which would not be enough for a good young queen.

The points mentioned may have taken some time to read, but when you have gained experience can be noted within seconds. The two most important points are whether the stock is queen-right and whether there is enough food on the combs.

FIRST YEAR BEGINNERS

By the end of the first week of July you will

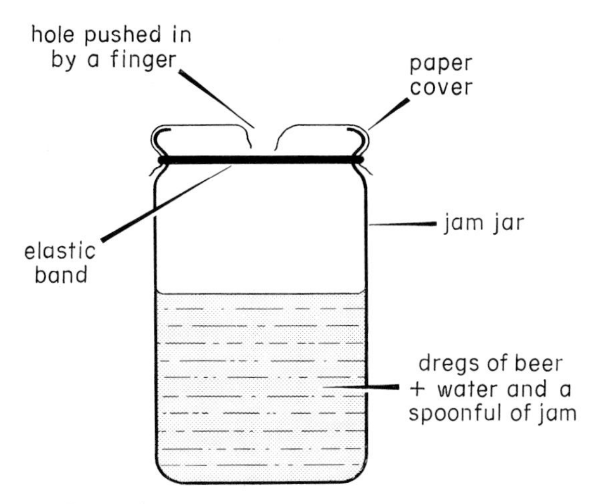

58 A simple wasp trap. Wasps can be a nuisance in August and September. One or two traps like this will divert them from the hives and reduce their numbers. Use no sugar or honey, though, or bees will also be trapped.

have had a month in which to examine your nucleus every few days and follow its natural expansion; the bees should be covering seven or eight combs now, with thousands of young ones due to emerge in the next few days, so I suggest that a queen-excluder be put over the brood box and a super over this, then the crown board. The shallow super with 11 frames of foundation will provide room for the bees, and in a normal year they will draw out at least five or six frames before the end of summer, storing perhaps 8–12 lb of honey in the middle three or four. In 1983, some nuclei had completely filled a super by the end of July, but this was unusual.

Work for July

a Assemble a super of 11 shallow (close-spaced) frames and fit with sheets of wired foundation;
b add this super over a queen-excluder during the first half of the month;
c go to apiary meetings and see how experienced beekeepers handle strong colonies, with more than one brood box and super;
d learn to 'read a comb' and identify brood in all stages, pollen and thin nectar as well as ripe honey. Using reading glasses if needed, make sure you can see eggs and spot recently polished empty cells ready for the queen to lay in;
e at the end of July restrict entrance for easier defence against wasps (and other bees).

August is the month when holidays are taken, and there is no essential bee work to be done so far as beginners are concerned; also bees can sometimes be less easy to handle for two or three weeks, but are usually normal again by the end of the month.

SUPERSEDURE

As experienced beekeepers know, this replacement of an old queen by a young daughter without swarming takes place much more often than is generally realized, especially towards the end of summer. The bees usually construct from one to four queen cells on the face of a brood comb rather than at the edges. If queens are not marked there is no visible evidence after a week or two, except perhaps the remains of one or two queen cells on the face of a central comb, usually almost completely torn down and only to be noticed by a very observant beekeeper. When a young queen emerges from a supersedure cell, there is seldom a conflict between mother and daughter, and sometimes both have been seen laying, occasionally even together on the same comb, for a week or two or even longer. In the end the bees tend to neglect the older queen and feed her less, so that she shrinks and becomes less and less attractive, secreting much less queen substance than her fine young daughter. Then one day she disappears, and if found dead outside the hive will not appear very different from a worker. This is the pitfall when re-queening,

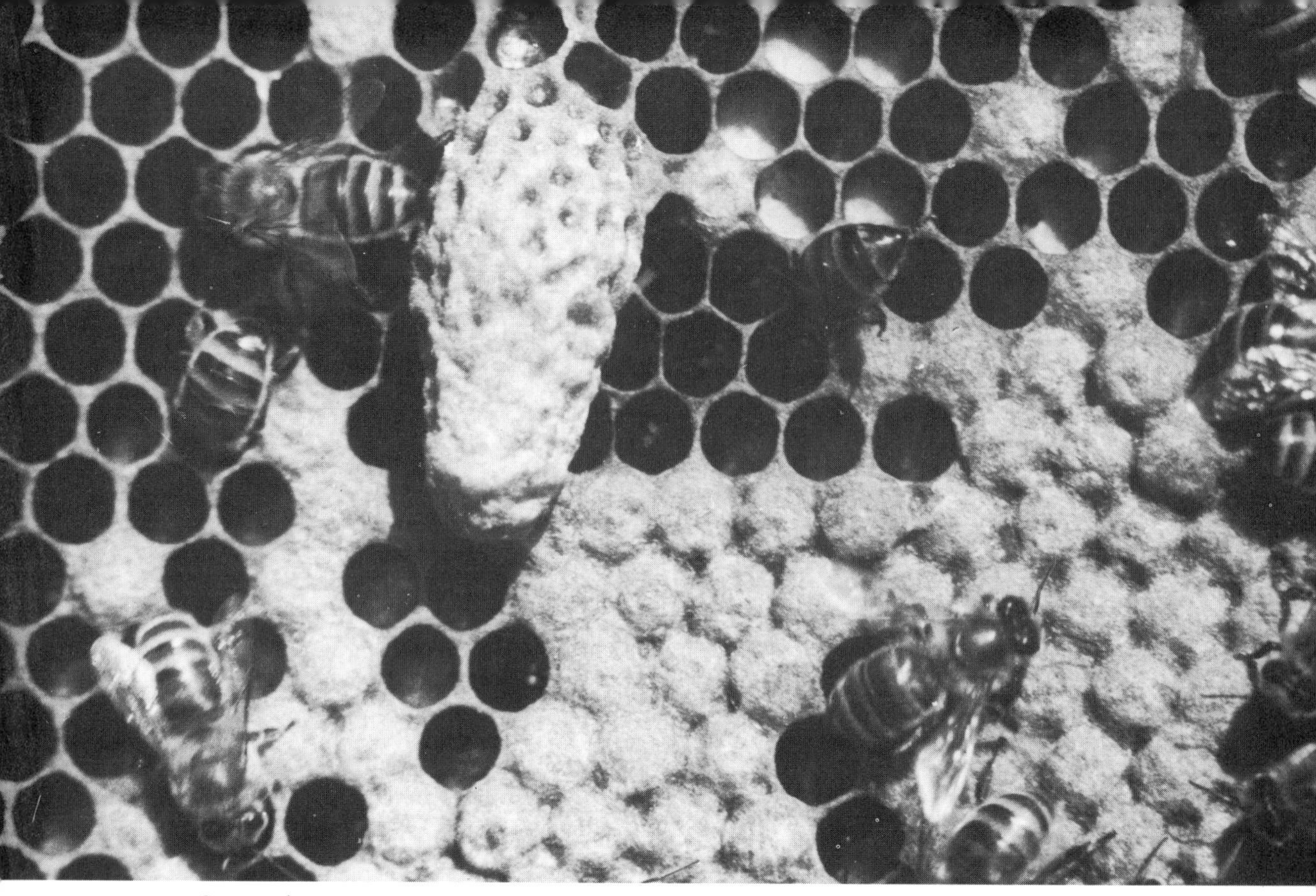

59 Supersedure queen cell, in centre of brood comb.

especially in July, August and September, and many a beekeeper has found and destroyed the old, marked queen and introduced an expensive young one, only to find her dead on the doorstep next morning, rejected by the bees who prefer the daughter they have raised themselves. So remember to look for evidence of supersedure before introducing that expensive new queen.

Some strains of bees will regularly supersede rather than swarm, but this can result in inbreeding, and after a few years this may produce 'pepperpot brood' caused by a large proportion of larvae being ejected by the workers, as they detect some abnormality. So it is good practice to introduce new blood every five years or so. Perhaps the easiest way of doing this is by giving a ripe queen cell without killing the old queen. The queen cell may be protected by a small strip of adhesive tape fastened around the sides, leaving the end free (see Fig. 24), by an orthodox queen-cell protector, or by an inch-long slice from a $\frac{1}{2}$-in diameter hosepipe. When queen cells are torn down by the bees, they are invariably attacked from the side, never from the end.

The protected queen cell may be gently pushed down between two central brood frames, leaving the upper end flush with the top bars and when the young virgin emerges she will normally be accepted as the daughter of the hive. This induced supersedure method of re-queening may appeal to beekeepers who experienced difficulty in finding the queen anyway. The success rate is normally about 80%, rather less than that with new queens introduced after killing the old queen. On the other hand, it is highly probable that the 20% failure rate is due to the fact that the bees are well satisfied with their present queen, and know more about her than the owner. There is an interesting field for some modest research here, by trying out induced supersedure not only on old queens, but also in a few cases on young queens in full lay. In any case, it is necessary that queens be marked, otherwise it is extremely difficult to be sure whether a colony has superseded or not.

Any deformity, abnormality or damage in a queen, whether visible to the beekeeper or not, can result in supersedure. Sometimes a much-travelled queen may have received rough handling and within a couple of weeks of introducing her the disappointed beekeeper may find queen cells and no queen. Sometimes a queen may be damaged during a routine inspection. I remember this happening many years ago, and later seeing a fine supersedure cell in the middle of a central brood comb, with the queen (still less than a year old) gallantly doing her best with a plainly dented abdomen, reminding one of a car body dented by a collision. Since that day, when replacing a frame with the queen on I have always waited until she was well clear of frame woodwork before putting it back into the hive. Although I have never tried this myself, I know from very experienced workers that supersedure can be induced by amputation of one antenna or a front leg. As a practice I would find this repugnant, but it is part of the body of knowledge on the subject.

HEATHER HONEY

Heather honey is the aristocrat of all the honies. Not only has it a noble appearance, with its dark, golden-amber glow, its jelly-like consistency, its aromatic bouquet and superb flavour, but also, with a high protein content, it is a wonderful food. Unique among honies in this respect, this high protein content (about 2%) stems mostly from the gelatinous, albuminoid nature of the honey itself, quite apart from the pollen content. Ling honey is thixotropic, i.e. when stirred it goes much more liquid for a short time before reverting to a stiff, jelly-like substance. When pure it never sets

60 Heather honey from Dartmoor, Devon.

hard, and I have specimens ten years old with no trace of crystallization. Even a small proportion of other honies (clover, bramble, fireweed) will cause fairly rapid granulation, although without affecting the flavour or food value of the honey – only its appearance.

True heather honey comes only from the ling, *Calluna vulgaris*, usually yielding for about a month from the second week of August (on Dartmoor). Scotland has the best areas, but there are also many in England and Wales which yield copiously when conditions are right. Quite often a number of beekeepers get together to borrow or hire a lorry and take 20 or 30 hives collectively up to 100 miles just to obtain some of this superb honey.

Young queen needed

Ideally, there should be a very large number of young worker bees in the hive, with more emerging during the first half of August; this means that the queen should be at her egg-laying peak over a month later than usual, and still laying heavily during the first two weeks of July. In practice this requires the colony to be headed by a young queen, mated earlier in the same year. Such a queen will be activating the bees with a maximum secretion of pheromones as well as keeping up the pressure in the restricted brood nest, whereas a slow queen housed in a large brood box will probably result in most of the honey being stored down below in the brood combs.

Single brood chamber

Another consideration is that eggs laid after mid-July will not produce foraging bees until late August, so that most beekeepers working on one and a half or double brood chambers will reduce to a single brood chamber around mid-July. An alternative method is to put the queen down in the lower brood box a few days before going to the heather, rearranging the brood combs so that this box has the sealed brood combs in the centre, with eggs and open brood on the outside frames. This ensures that the outside combs are occupied by brood for longer, so that the tendency of the bees to store honey in the outside brood combs is prevented.

Heather supers

Because it is gelatinous (technically thixotropic), or like thick marmalade, heather honey cannot be extracted from the combs in the usual way by centrifuging. Instead, most people either go for honey in the comb, which readily sells in 8-oz plastic boxes, or else press out the honey. If it is intended to produce honey in the comb, then unwired foundation is normally fixed in the shallow super frames. For the very best combs, with no midrib at all discernable as one bites through, small starter strips of thin wax foundation are often used, but it helps if these can be at least partly drawn out into comb before going to the heather. A frame of drawn comb placed at each end of the super encourages bees to cluster and draw new wax. For liquid honey, a super previously used for the normal summer honey flow can be placed above the brood box, over a queen-excluder, while still wet from the extractor. This has the advantage of attracting the bees into it very readily, but on the other hand the admixture of even a small quantity of other honies can cause granulation. To avoid this a 'dry' super can be used, i.e. a super of extracted combs placed on a hive and licked dry by the bees. Of course the heather honey has then to be pressed out, thus destroying the combs which might otherwise have been used over and over again for normal floral honey. However, the wax can be recycled back into foundation and it is good practice to renew some combs every year to prevent too much darkening and embrittlement with age.

Final preparation

The object is to crowd as many young bees as possible into a hive comprised of one brood chamber and one (possibly two) supers. In the last days of July this is normally achieved by putting clearer boards under the summer honey supers, to get the bees down into the heather super. If the super has thin starter strips of wax, then a feed of a gallon of thick sugar syrup will get them drawing out the wax into combs.

Site at the heather

It is usually possible to arrange a site by agreement with a farmer or with the forest authorities, but alternatively one can drive around with a one-inch Ordnance Survey map and take a calculated risk by putting a couple of hives in an isolated gulley or hidden spot not too far from the road; invisibility is still one of the best security precautions. For the cautious, it is probably best to call on and chat up the occupants of a cottage or farm and strike a

61 Hive roped up for transit (note ventilated roof). One end of rope has a loop, through which the other end is threaded. This gives a 2 : 1 leverage for tight roping.

bargain, softening up with a pot of your summer honey and the promise of heather honey to come.

Moving the bees

Personally I just block up the hive entrance with strips of foam rubber at 4.30 a.m., rope up securely (see Fig. 61) and drive straight to my site in Dartmoor, as this takes me only an hour and in the cool of the morning the bees take no harm. For a longer trip, or for later in the day, it is necessary to replace the normal cover board with a wire gauze screen to give ventilation, as bees confined for several hours under crowded conditions can get overheated and perish in a mass of softened wax combs and warm honey. Within the car boot or trailer, the hives should be placed so that the combs are lengthwise parallel to the road, to minimize frame movement when the vehicle is in motion. If the journey is going to take some hours, take a bottle of water and sprinkle half a cupful over the screen board every hour. By fanning and evaporating this water the bees will keep cool. At the site get the hives as level as possible on the ground and push an old piece of hardboard under the front to keep an area clear for the bees, otherwise in a couple of weeks grass or bracken may almost block the entrance.

On the moor

Sometimes it happens that the weather changes to cold, wet and windy a couple of days after the hives have been moved. It has to be remembered that, with crowded stocks having brood to feed, food can quickly be exhausted. One can either feed a gallon of thick syrup before the move, or have a Miller feeder on each hive so that if necessary a quick visit with jerrycans of sugar syrup can be made. A good alternative is to make up the brood chamber with one flank comb solid with food, and probably as much again in the corners of the other ten frames. Usually in three or four fine days the bees will gather enough food to see them through almost any bad weather after that. Ideal conditions are calm, warm and moist; on Dartmoor we notice that if the wind changes to east or north-east (cool and dry) the heather flow is turned off as if by a tap. This is when the stocks get bad-tempered, and even the gentlest of bees will sting for no apparent reason.

Although at times fickle, at its best the yield on the heather can be terrific, with two supers filled in ten days. Recent years on Dartmoor have not been so good, and a fair surplus of at least one full super has been obtained only in one year out of two. However, the brood boxes are heavily stored nine years out of ten, and little winter food is then required. Another bonus is the excellent spring build-up usually achieved by stocks which have been to the heather the previous August. Probably the high protein content of heather honey accounts for this, helped by a young queen stimulated into more intensive egg-laying by the August/September flow, to produce young bees to go into winter in showroom condition with very few hours in their log-books.

Extracting and packing

There are specialized techniques involving apparatus with perforating needles to pierce the cell bases and liquify the thixotropic honey by agitation so that it may be extracted centrifugally, but most beekeepers work for cut comb and then press the combs remaining, either in an MG heather press (see Fig. 67) or in a home-made model. The method of cutting and packing 8-oz combs has already been described.

If stirred and agitated, the pressed honey will pass readily through the normal filter on a honey tank and may be bottled immediately. The jar should be filled to the top, as the air bubbles characteristic of pressed heather honey plus the slightly higher water content make it weigh less than ordinary honey, so that it can be difficult to get a full pound into a normal jar. Also for these same two reasons, heather honey is more liable to ferment than most honies. This can be prevented by warming the open jars of honey, about two dozen a time, on the rack of a fridge warming cabinet, illustrated in Fig. 84. After 6–8 hours the honey temperature will be up to about 120°F (50°C) and effectively pasteurized; for the benefit of professional nutritionists who may read this, although 120°F is somewhat below the temperature range normally quoted for pasteurization, with heather honey it has been found effective, and of course no honey should be heated more than absolutely necessary. By surface evaporation an invisible 'skin' of denser honey is formed on the surface and this acts as a barrier against subsequent aerobic fermentation. The surface resembles that of a jar of well-set home-made marmalade, and a jar may be inverted without honey running out.

Finally, remember that heather honey normally sells at a premium of 25–40% over ordinary honey, and is much sought after by people who appreciate the finer things in life.

Six

Autumn – the annual harvest

HARVESTING THE CROP – BEGINNERS

By now (end of August) your original four-frame nucleus should be filling most of the brood box, with a fair cluster of bees in the super you put on in July. With a nucleus made before mid-June, if you are fortunate, there may even be a few frames of sealed honey. The purists may say that this should be left for the bees, but the master is also worthy of his reward and you need have no qualms about taking the six to twelve pounds of honey for yourself: this will help to maintain the interest of your family, and of your neighbours, which in the long term helps the bees.

Clearing a super

There are two simple ways of getting the bees off the combs: by brushing or by using a clearer board. As only four or five frames are involved, it should be fairly easy to brush them off. Go to work as described previously, and when you have the roof off, push the hive tool gently between the super and the excluder and puff smoke into the gap before levering up the super; at this stage make sure that the excluder is not also being lifted, and lever it down gently to break contact. Finally, lift off the super with a slight twist and place it diagonally on the inverted roof, close to the hive. Put your cover cloth over the main hive and come back to the super. Just by looking down into it you will be able to see which frames have been stored in, and these have to be removed, placed in another box and covered with a cloth to prevent other bees finding them. As you take out each frame, hold it over the hive and brush off the bees with a firm downward motion, using a large feather, a pigeon's wing or a soft nylon brush obtainable from your local agent, partly unrolling the cover cloth so that the bees may get back into the hive. Check that at least half of the honey is capped, i.e. that the cells are covered with wax, and also check that no honey falls out of the open cells when the frame is held horizontally over the hive. Quite possibly there will be one or two combs containing nectar (which falls out) rather than ripe honey, and these should be left in the super which should be replaced on the hive, over the excluder. Finally, replace the crown board and hive roof, and take away the box of frames containing your first crop of honey.

At this stage you may not possess your own extractor, but your branch will most likely have one which members can arrange to borrow for 48 hours; alternatively a friend may agree to extract your frames for you, or even come along just once and show you how it is done. Don't expect this favour more than once in your beekeeping career, however, or you may develop into a bee-owner rather than a bee-keeper! Briefly, the method is to slice off the wax capping from both sides of the combs with a long, thin knife, and then put them into a small centrifugal extractor which works like a spin-drier, except it doesn't spin so fast. The cappings may be drained in a colander to give a little more honey, and then put on a plate on the crown board under the roof, for the bees to lick dry before melting down into a small cake of beeswax. The wet, extracted combs should be put back in the super on the hive as soon as possible, taking care to place any combs containing cells of pollen in the centre. Make sure that crown board and roof are replaced to fit exactly, so that no wasps (or other bees) can get in to rob; for the same reason make sure that the front block is in, to restrict the entrance which has to be defended.

A nucleus with a young queen reaches its peak much later than an older stock, so your queen will probably continue laying well into September. Encourage this by continuing to feed, so that there will be plenty of young bees to see the colony through the winter safely. At this time of year use a thicker syrup, containing 2 lb *white* sugar to 1 pt water. The point here is that syrup fed later in autumn is mostly stored and capped for use in winter and spring, and dilute syrup involves unnecessary labour by the bees in evaporating it down to the required honey-like consistency. There is no exact guide to the amount of sugar to be fed, as the bees will also be gathering and storing honey for themselves, from late Michaelmas-daisies and (in October) from ivy, but a good rule is to carry on feeding thick syrup up to the second week in October, so long as the bees take it down. In the northern half of Britain feeding should be completed two or three weeks earlier.

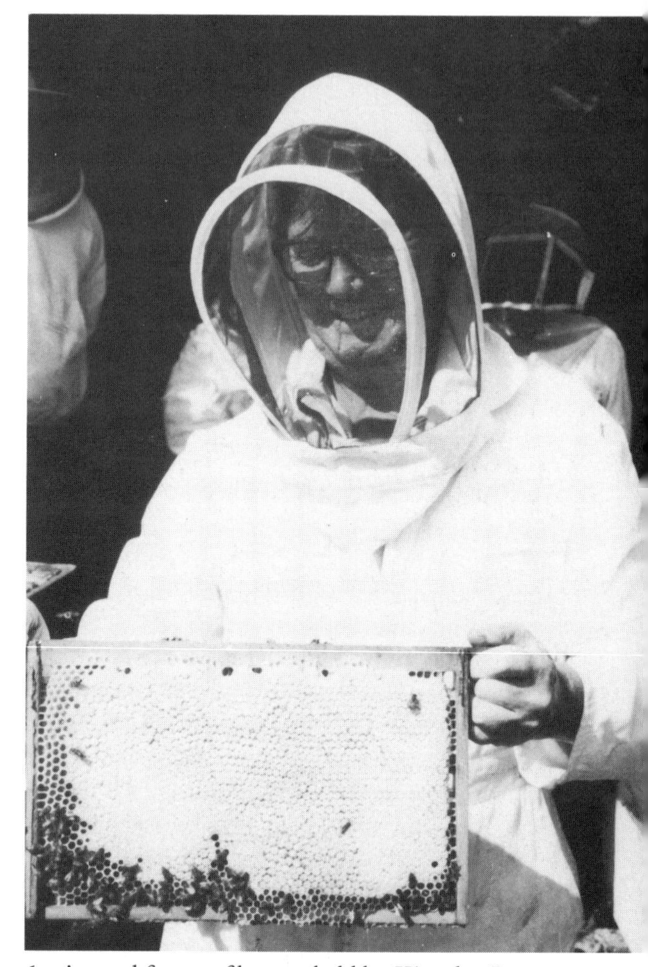

62 A good frame of honey, held by Kingsley Law, editor of *Beekeeping* magazine.

HARVESTING THE CROP – IMPROVERS

When?

Unless one is taking bees to the moors for heather honey, there is not usually much nectar available to bees in August, except for the last of the bramble and rose-bay willow-herb (fireweed). In most of Britain the main honey flow is over by the first week in August, or even the end of July, and some people like to extract the honey crop immediately. However, there are three good reasons why this is not ideal:

a there may not have been time for all the honey to have been fully ripened and sealed over; unripe honey is thin, very runny and does not store well;
b the bees are often bad-tempered for a couple of weeks at the end of the honey flow;
c if all the honey supers are taken off so quickly, there may be a very large worker force without enough food, or even enough room to live, and while the beekeeper is away for the annual August holiday the bees may starve if the weather should be really bad. It is better to wait until the last week in August or even the first week in September, when the bees are much better-tempered, there are fewer of them and the honey is well ripened.

Taking off the supers

If one just has a couple of hives in the garden, the best way to get the honey off is to slip a clearer board under the supers one evening and lift the supers off the following evening or the day after. Porter bee-escapes, which allow the bees to get through the fine springs one way only, are made to fit into the normal feed holes on crown boards, so that it is easy to lift the honey supers and slip the board between them and the brood box. (See Fig. 63.) It is easier still if one person lifts the boxes just a few inches while another slips in the board.

63 Porter escape, showing springs. The upper plate with rounded ends slides on lower plate bearing two pairs of fine springs. A bee can enter through the round hole and push between the springs, but cannot return. Crown boards fitted with Porter escapes and placed between super and brood nest will clear super of bees in 24 hours.

Some years ago the writer arrived at an out-apiary 14 miles away to find that the Porter bee-escapes had been left at home. Fortunately all hives in this apiary were fitted with home-made crown-board feeders (Buckfast type), and these were tried as clearer boards. They worked perfectly, with the bonus that they were in the right place for winter when the supers came off two days later.

A simple temporary replacement for the crown board on the top is an 18-in square of old carpet. If the supers are at all old and ill-fitting, it is most important to check that there are no gaps where wasps (or bees) can get in to rob. A roll of wide masking tape is useful here, to run around the box-to-box junctions.

If the hives are at a distance, it may be inconvenient to make the two journeys necessary when using escape boards. On a small scale it is usually quite practicable to shake, bump or brush the bees off each individual comb and stack the frames in a spare super, covering with a cloth. There is a special soft bee brush sold for the purpose, but a goose feather, pigeon's wing or even a handful of grass can be used. Where supers are incompletely filled, only the full frames should be taken, and the odd two or three (at the sides, most likely) containing only a little honey used to make up a super of such combs to put back over the brood box, later to become the reservoir of winter food.

On a larger scale, a bee-repellant like benzaldehyde (synthetic oil of almonds) may be used. A few drops of this sprinkled on a cloth and laid over the top box will send the bees down into the brood box. It is best to give a puff or two of smoke first to start the bees on their way down. Keep the cloth stored in a long tin (Dr Oliver's Bath Biscuits?) to avoid evaporation between apiaries. It is usually convenient to use two such cloths and transfer them from hive to hive, so that one is working while the other is being moved. About two to four minutes is usually enough, but it works more rapidly on a warm day, and while one super is being lifted off and loaded on a vehicle the next is being cleared. On a cold day it may not work at all, but another substance (butyric anhydride or 'bee-go') does.

Anti-crush wedge

Often, when splitting heavily propolised boxes by inserting a hive tool at first one corner and then another, it is difficult to prevent the upper boxes from sinking down again, closing the narrow gap and killing bees in the process. A useful device is a small wooden wedge on a piece of string tied to one's overalls, always there to keep the gap open and prevent the crushing of bees.

Avoiding back problems

Supers full of honey can be very heavy, and it is important to learn correct lifting techniques so as to avoid a painful 'beekeeper's back'. The human spine is not designed to carry loads when bent, and the secret is to make the leg muscles do the lifting, not the body muscles. Much lifting can be avoided, for example by not putting heavy supers down to ground-level in the first place. When putting a clearer board on a hive single-handed, the supers can be temporarily moved across and placed on the roof of the next hive, over a narrow eke to avoid crushing the bees. If a heavy super has to be carried some distance, then lodge the near end on your hip and grasp the far end at the base, with your arm going over the super. This way the wrist takes only half the load and the spine stays erect.

Honey extracting

There is no need to buy a large, power-driven extractor and for up to 20 hives a second-hand, manual six-frame tangential extractor is perfectly satisfactory. These can be obtained from sales in the classified-ads columns of the bee journals, but most bee clubs have one or two for loan to members. There will probably be a club within a few miles of you, and you should belong to it anyway. For example, Devon has 13 such clubs and it would be difficult to live more than 12 miles from one of them.

There are purpose-made uncapping knives, some electrically heated, and also heated un-

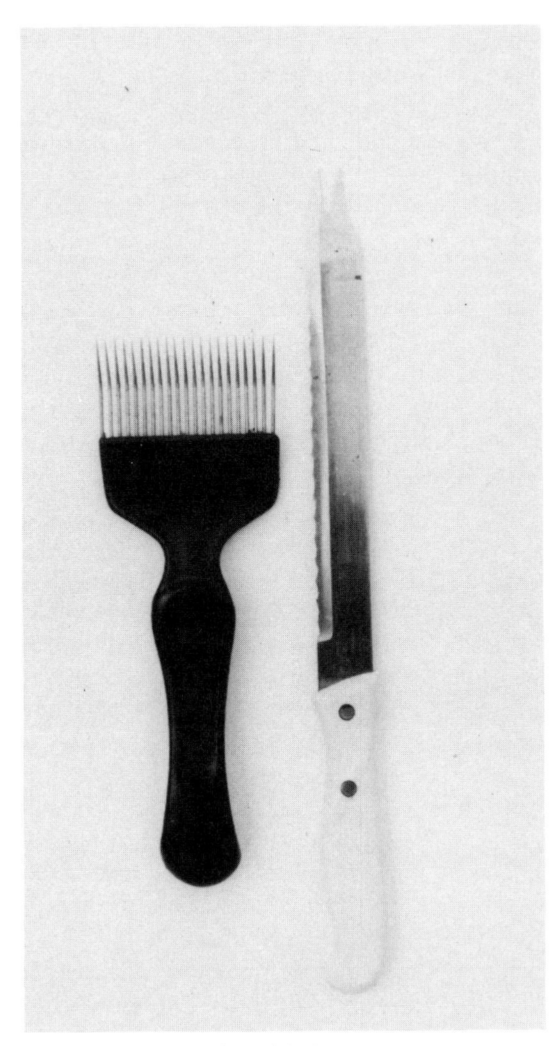

64 Uncapping knife and fork.

capping trays, but these tend to overheat the honey and may give it a slightly caramelized flavour. Worse, they may reduce the diastase (enzyme) content. The writer has tried all these aids and came back years ago to a thin, sharp, wavy-edged Kitchen Devil bread-knife with a 7-in blade, used in conjunction with a purpose-made uncapping fork, $2\frac{3}{4}$ in wide with 21 tines. The former is sold at fairs and shows (no doubt in hardware shops, too); the latter by all the bee-appliance dealers. In practice, one works over a sieve resting on a two-gallon plastic bucket. The actual motion is similar to that of an expert chef carving a large ham, slicing neatly under the wax capping to leave an exposed surface of open cells full of honey. With experience one acquires what can only be described as a three-way attack, combining a gentle sawing action with a yawing motion and at the same time taking the knife along the length of the frame. If there is a depression on the comb surface, which often occurs in the lower central area of combs from the middle of the super, then put down the knife, pick up the fork and with a somewhat similar motion run under the cappings of the lower cells, which the longer blade of the knife cannot easily deal with. Both tools should be scraped clear of honey and wax on the side of the sieve. When the sieve appears to be full and clogging, stir up the wet cappings with a stiff kitchen knife (round-ended) and take a break to let the honey drain through. If in a hurry, invert the sieve of wet cappings over another plastic bucket; the whole mass will fall away in about half a minute, and one can start again. Later on, the accumulated wet cappings can be pressed, or better still centrifuged in a spin-drier or purpose-made cappings spinner, to get out most of the remaining honey, before making mead from honey still left on the wax.

As each frame is uncapped, stand it in the wire cage of the extractor. When full, rotate slowly at first, and build up to a very moderate speed, for half a minute. Then stop, reverse the combs and once again rotate slowly, but build up to a fast speed and hold it until the honey is no longer heard pattering on the inner side of the extractor (one minute). Finally reverse again and run fast for another minute. Unless these precautions are taken the full weight of honey can fracture the combs when rotating at high speed. If about half the honey on one side is spun out first, then the same on the other side, one may use a high speed to get out the maximum possible amount of honey. It is important that the combs should be warm, as cold honey is much stiffer and more difficult to extract. Combs straight from the hive are usually warm enough, but boxes of full combs stored overnight should be in a warm place, with a couple of blankets over. If late in the season, it may be worth while to stack the full honey supers, eight at a time, for 48 hours over a 60-watt bulb in an empty deep box, with a 'bridge' of sheet tin or aluminium over the bulb (and two inches above it) to disperse the heat and prevent a 'hot spot' on the combs immediately above. Cover the pile with insulation, like a polystyrene square, with old blankets draped over. It also helps to have a warm room to work in. As a matter of interest, the optimum temperature for uncapping and extracting is 70–80°F. Above 90°F the wax softens and 'drags' on the knife. Below 65°F, the honey is stiff and too much is left behind. After warming as described for 48 hours the honey temperature will vary from about 68–70°F in the top super to 78–82°F in the lowest one, just above the lamp. When the level of honey in the extractor has risen so that the frame lugs are touching the surface and dragging, take a break and run off the 40–50 lb honey a six-frame extractor will then contain.

A 56-lb stainless steel honey tank is invaluable here, especially if it fits under the honey valve of the extractor when the latter is operated on a sturdy old kitchen table. It helps to have a three-tier filter system to avoid clogging and waste of time. The simplest and most effective method is to slip a kitchen colander into the top of the tank so that any relatively large wax fragments are intercepted first. The built-in filter mesh of the tank top takes out most of the remaining fragments, and a fine nylon cloth (wine filter bag from chemists) tied under the tank mesh will take out any very small particles. Honey drained from cappings can be tipped into the tank top at the same time, and the honey bottled after standing for 24 hours to allow bubbles and froth to rise to the top. The 56-lb tank will supply honey for immediate bottling and the excess can be run directly (unfiltered) into the standard 28-lb honey tins for bulk storage or sale, preferably lining the tins with large plastic bags. Some beekeepers bottle straight from the tank, using only a coarse, conical filter; although this honey will not win a prize at the local show, it is very good to eat. In fact, steadily increasing numbers of customers ask for 'natural' unfiltered honey, and appreciate the value of fragments of propolis, pollen grains, even wax as a health food. I combine virtue with labour-saving and cheerfully say that no extra charge is made for this 'special' natural honey, sold with distinctive red caps on the jars!

Honey jars

New jars from the factory are packed in half-gross lots; they look clean but are not, and must be washed. With practice it takes very little time to wash them in a plastic basin of very hot water with a squeeze of detergent, using a washing-up brush. Drain, then rinse in clean, cold water and drain again, finally reverse on trays and leave in a warm room to dry naturally, i.e. not wiped with a cloth.

Honey for sale to the public should be check-weighed on an accurate scale, or an extra quarter-ounce allowed to make sure no jars are underweight. Honey for personal use or Christmas presents can be stored in a variety of containers.

65 Warming honey before extraction. Six to eight supers are warmed for 48 hours. Criss-cross the supers so that the frames are at right-angles to those above and below.

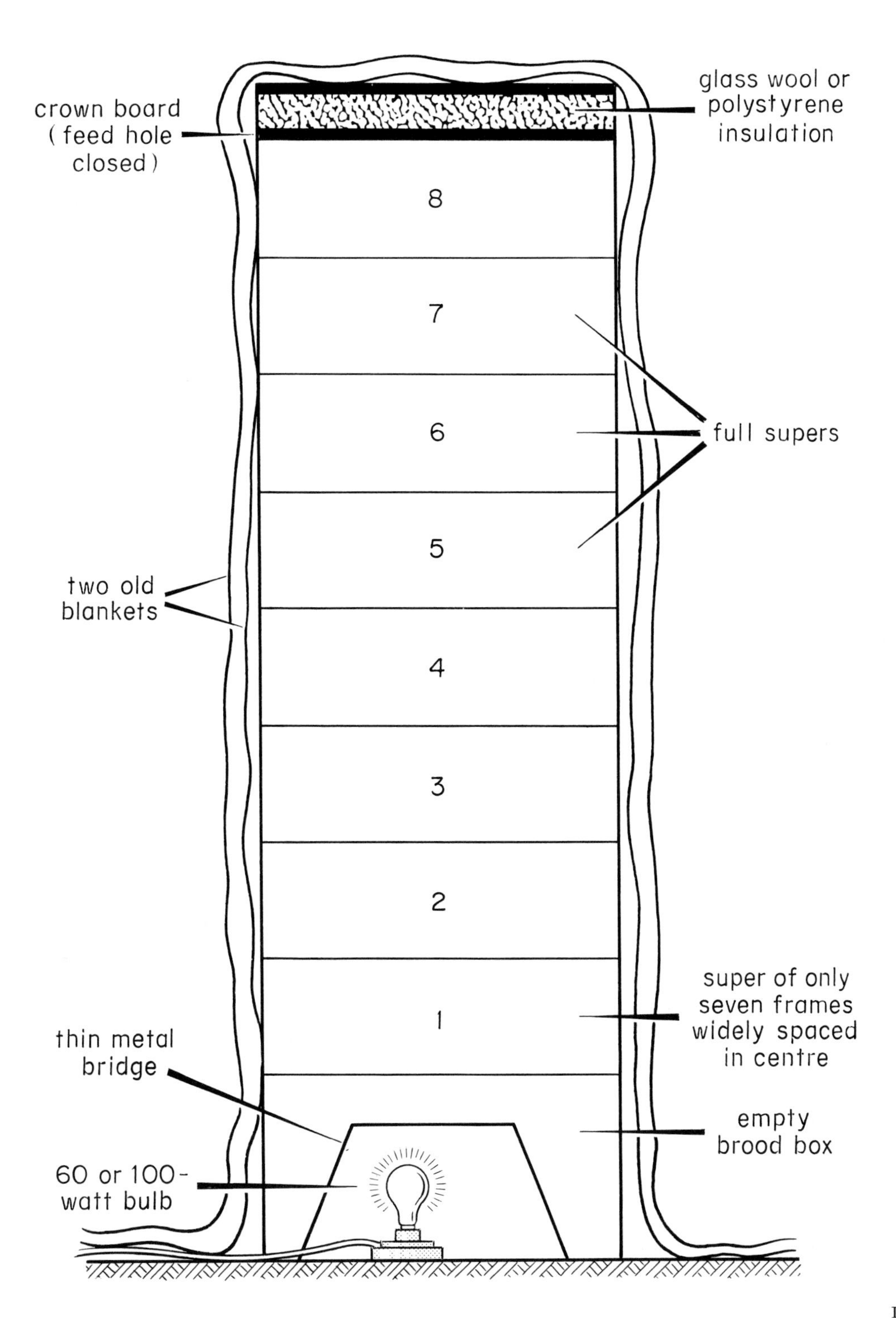
crown board
(feed hole
closed)
glass wool or
polystyrene
insulation
8
7
6
5
4
3
2
1
full supers
two old
blankets
super of only
seven frames
widely spaced
in centre
thin metal
bridge
empty
brood box
60 or 100-
watt bulb

QUEEN INTRODUCTION

Why re-queen?

There are many situations in which it is necessary to introduce a new queen. The old queen may be failing and producing far too many drones, or nothing but drone brood, even in worker cells. The bees may be too bad-tempered to be tolerated in a garden at home. A swarm may be headed by a queen believed to be old and unlikely to get through another winter, or perhaps a change of blood is thought desirable to maintain stock vigour and resistance to some virus disease. In all these situations it is necessary to remove the old queen first, and this must be done before the new queen is introduced in her cage. Queens may be introduced at any time between March and October if necessary, but in my opinion the end of August or early September is as good a time as any, as a routine.

General principles

As always, it pays to go with the bees and to give them what they are expecting. For example, a stock which has lost its queen a week ago will be trying to raise its own and will readily accept a ripe queen cell or even a recently emerged virgin queen. A stock queenless for some weeks will be suspicious of a mated queen which suddenly appears in their midst and may not accept her, but will do so more readily if given a frame of young brood and eggs a few days before the new queen. Young bees will accept a new queen more readily than will old bees, and a hungry queen soliciting food and not rushing arrogantly about, will be more welcome. Queens are also accepted more readily during a honey flow, hence the advantage of feeding the hive for a couple of days if no nectar is coming in.

Occasionally one may remove the old, marked queen at a time when she has already been superseded and there is a young daughter also there, perhaps even laying. In this case the new queen will be rejected, possibly without the beekeeper being aware of anything, unless perhaps a young dark queen is noticed weeks later when a fair queen had been introduced.

Caged queen by post

A queen arriving by post will do so in a small travelling cage provided with a plug of candy and accompanied by eight to twelve worker bees. Queens in such a cage with attendants may be kept at room temperature for several days if necessary, but two drops of water on the gauze every day would be appreciated. Cages may also be stored over the crown-board gauze of a hive for a couple of weeks or kept in a 'queen bank', as already described. It is possible to push the cage directly into a *queenless* hive, between a couple of brood frames, after removing any cork or adhesive tape covering the candy plug, but better results are obtained by removing the workers first. This helps because there are then no alien workers likely to antagonize the bees in the hive, thus allowing more ready acceptance of the new queen. When removing attendant workers from cages it is convenient to sit at a table in front of a closed window, in case the queen decides to fly. Usually the staple pinning down one end of the wire gauze panel can be lifted with the blade of a penknife or a fine screwdriver, and the workers gently flicked off as they can be reached, often without the queen leaving. Sometimes the queen and workers sit tight, in which case shake them gently on to a soft cloth and immediately scoop up the queen and press back the staple fastening the gauze. I have often done this in a car with windows closed (and car heating ventilator slots blocked with tissue paper!)

There are some advantages in transferring the queen to a plastic hair-curler cage having a

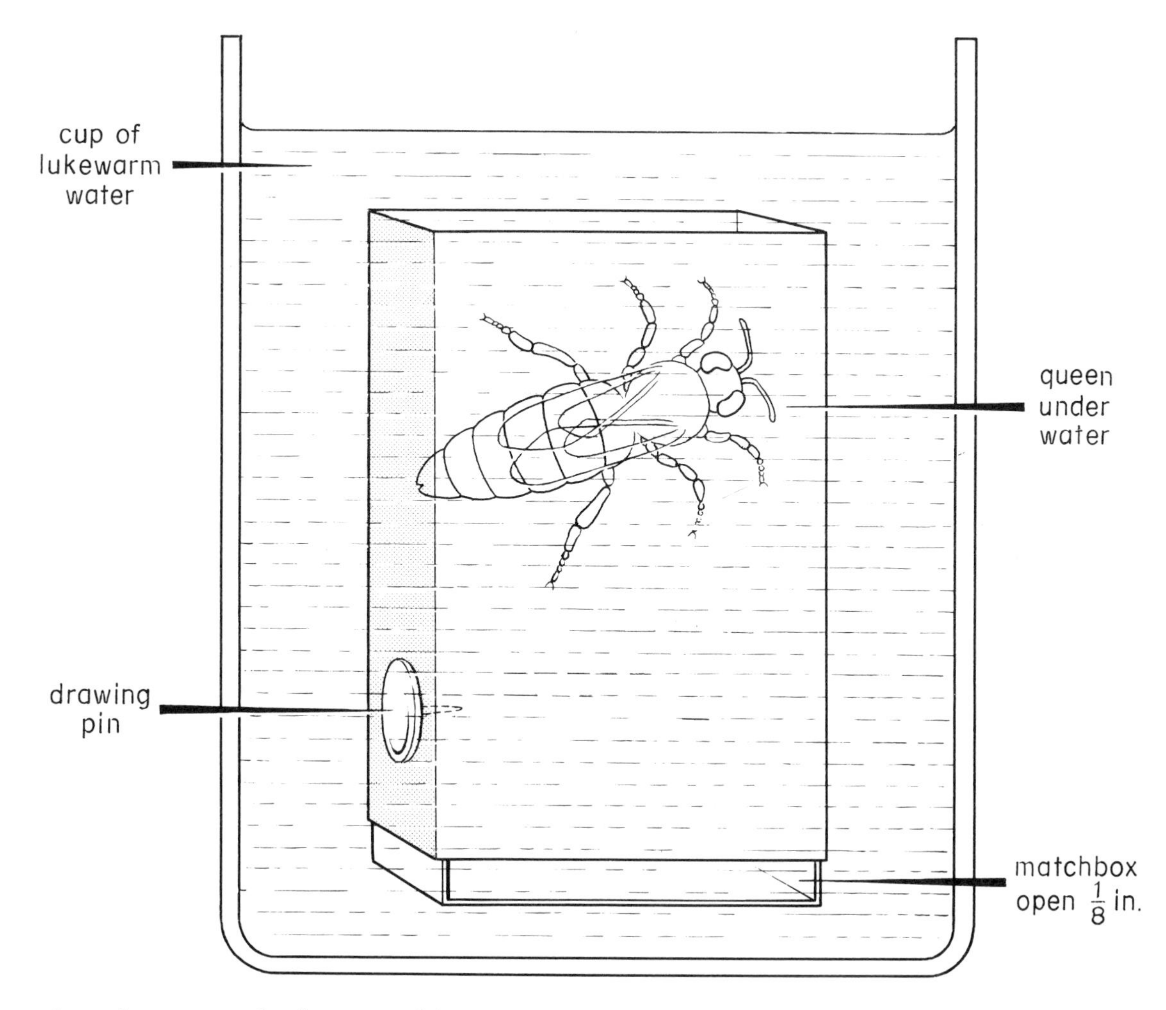

66 Queen introduction – the water method.

plug of queen candy (honey and icing sugar worked to a stiff putty-like consistency) one end and a cork the other. The flexibility of such a cage enables it to be pushed between frames more easily and the all-round access by bees enables them to feed the queen and even to lick her. If the wooden travelling cage is used, make sure that the wire gauze panel is more or less horizontal and underneath, so that bees in the hive can get their tongues through to feed the queen.

The water method (Snelgrove)

Keep the queen by herself in a matchbox in your pocket for an hour without food, with the drawer very slightly open for ventilation (and a drawing pin through the side to prevent accidental opening or closing). De-queen the stock concerned and also destroy any queen cells present. Use a cup of lukewarm water (at body temperature) and push the matchbox completely under so that it fills with water, hold it there for six or seven seconds, lift it out so that the water drains away, open and invert over the feed hole, or drop the wet queen into the hive between the brood frames. A hungry, humbled queen wet from her bath is accepted under almost any circumstances, whether she be mated or not, and the workers lick her and feed her without question.

Laying workers

One certain way to solve the problem of any queenless stock, even one with laying workers which may have already refused to accept a new queen, is to shoot in a swarm. First smoke the stock, take out two of the central frames with bees on them and shake the swarm into the empty space. Then shake each frame of the original bees back into the hive and gently replace the frames.

Miscellaneous

A weak stock is always more anxious to welcome a new queen than a very strong stock, so that a valuable breeder queen should always be introduced to a small nucleus in the first instance and subsequently united to the main stock.

When a stock is de-queened, the bees usually discover their loss in 10 to 25 minutes and will be seen running around at the entrance and up and down the front of the hive in an agitated manner. A new queen is more readily accepted now than a couple of days later, when they will have begun to build queen cells.

For rapid re-queening, an alternative to the water method is to remove two brood combs, shake gently over the hive to get rid of older bees while retaining younger bees (which cling on more tightly), and place temporarily in a box or upturned roof. Drop the new queen into the young bees between these two combs and replace in the hive.

In autumn, any stock either made queenless or formed by shaking or brushing surplus workers out of supers will take a queen with little argument. The writer has often re-queened in September by pushing the candy-plugged end of a hair-curler cage containing a queen into the hive entrance at about 5 or 6 p.m.; within ten minutes a 'beard of bees' will be covering the cage and two hours later she will have been liberated. By nightfall, only an empty cage remains, with a contented humming in the hive.

Direct uniting

It is also possible to unite small stocks directly as a method of re-queening, and the secret of success here is exposure of combs to light and air, taking three or four minutes before pushing the different combs close together in one brood box, after puffing smoke down between the frames.

Whatever method you use, it is wise to leave the colony alone for about a week, and even then do not make a close examination of the combs but be content to remove the empty cage. If by any mischance the queen has not been released and the candy plug is not eaten, bend back the gauze and drop her in. If the candy plug has been eaten but the queen is still taking refuge inside, take out the cage and queen and close up the hive; it is likely that a previously unsuspected young queen is there already.

MEAD FROM CAPPINGS

In 1797 the secretary of the Western Apiarian Society, the Revd Jacob Isaac of Moreton-hampstead, gave this recipe for mead. 'Take 1 lb honey per quart of water, boil for 15 minutes and scum off. Boil again with 1 oz fine hops per gallon and run through a fine linen surplice. Cool and tun. Three weeks later bung down. Bottle at 12 months and drink after 3 years.' No mention of yeast, but no doubt the same 'tun' or cask was used year after year and retained enough natural yeast to start up fermentation. We shall never know how our mead today compares with Devon mead nearly 200 years ago, but here is the method which I have used for some years, utilizing honey which cannot otherwise be harvested.

When extracting for the year has been

67 A heather honey press.

completed, there is usually a bucketful or two of wax cappings left over, containing much more honey than is commonly supposed. After draining, pressing and even centrifuging, the apparently dry wax still contains over 50% by weight of honey. If fed back to the bees they will get most of this out, but only if the cappings are spread out thinly, probably involving more than one hive. They get excited in the process and much of the honey is thus used up, so that making mead is perhaps a better way of using the honey that cannot be separated from the wax.

Washing the wax

Take a large plastic bin with $1\frac{1}{4}$ gal. cold water and tip in about 6 lb pressed or centrifuged cappings (only $4\frac{1}{2}$ lb if not pressed). Roll up your sleeves and crumble the wax under the water with your fingers until it is broken into very small pieces, so that the water can get at all the particles to dissolve the honey sticking to them. Allow the mixture to stand for half an hour and stir again with hands or a long spoon. Now push down a colander or small sieve so that a reasonably clear liquid comes through, and take out about a pint by dipping with a tea cup. Test the specific gravity of this with a hydrometer, or by floating a fresh egg. An average strength, neither particularly dry nor sweet, will arise from a starting gravity of about 1.105, when an egg will float with a patch the size of a £1 coin above the surface. A lower specific gravity, say 1.080, will produce a dry mead suitable for use as a table wine, and higher gravity, 1.130, will produce a sweet mead. The specific gravity of your honey water or 'must', as it is called, can be adjusted by adding water, or another handful of cappings and stirring.

Preparing the must

The next stage is to separate out the wax from the must. This can be done by first pouring through a coarse household sieve while holding back the wax with the fingers. Then take up and squeeze the wax fragments between your hands into balls the size of an orange; they will still hold a great of moisture, but not much honey. Later on, the wax can be spread out on an old cloth to dry, before being melted down into moulds for storage. The yield of must will be about nine pints.

The next operation is to boil the must (in an aluminium, enamel or stainless-steel vessel – not brass or copper). This brings to the surface pollen grains, wax, propolis, wood splinters and miscellaneous impurities which can be skimmed off. It also kills unwanted micro-organisms and sterilizes the must ready for fermentation by the yeast of your choice. Prolonged boiling is neither necessary nor desirable; just bring to the boil and switch off. Some mead makers prefer not to boil at all but

68 Making mead.

to sterilize with sulphur dioxide (Campden tablets plus tartaric acid), but on balance it is probably preferable to boil. When cool enough to handle, you can filter through a 'fine linen surplice' as our reverend predecessor did, or through a well-washed old sheet or shirt, into a narrow-necked gallon fermenting flask and a spare wine bottle. At this stage a large air space is needed above the must – hence the extra bottle.

The fermentation

Two or three days previously the yeast (preferably Maury, but any good wine yeast will serve) should have been activated in a pint milk bottle half filled with warm water plus a dessertspoon of honey and a squeeze of lemon juice. For protection, cover with a small plastic bag and a tight rubber band. Assuming that the fermentation is going well in the starter bottle, the contents should be added to the must while the latter is still warm to the touch. Also add the juice of two lemons (or a small teaspoon of citric acid crystals), a couple of wine nutrient tablets and a cup of strong tea (as left in the pot and discovered at washing-up time). Again protect with a plastic bag and rubber band. Without going into great detail, any fermenting liquor needs acid and some nourishment for the yeast, also tannin to give a slight astringency and to assist natural clearance. Only grapes (and elderberries) provide all these naturally. Brother Adam adds no yeast nutrient, but then he makes mead from heather honey (washed from pressed wax combs), and this honey is unique in containing up to two per cent protein.

There is usually a strong fermentation for two or three days, and when this has subsided the contents of the wine bottle may be added to the main liquor, leaving just a small air space, and the flask fitted with a rubber bung and 'glopper' (fermentation lock). After three to five weeks at room temperature (not less than 65°F) the fermentation will almost have ceased, and after standing for 24 hours in a cooler place (unheated room, garage shelf) the mead can be *siphoned* (not poured) off the sediment of yeast cells into another vessel, temporarily, and the gallon flask rinsed out before the mead is poured back and corked. After a further six months another slight sediment will be noted, with clear liquid above. At this stage siphon into clean wine bottles, cork securely and store horizontally in a cool place for at least another six months before drinking. As the Revd Isaac said, it is better still left for three years.

Note

In a carefully recorded extraction run some years ago, 62 lb of drained cappings from 16 supers produced 23 lb of pressed cappings apparently almost dry. From this four gallons of mead were made and just nine and a half pounds of dry beeswax obtained. For the benefit of non-beekeepers who have no access to wax cappings, a gallon of medium-sweet mead for general use may be made from three and a half pounds of honey. A sweet mead will need half a pound more and a dry mead half a pound less than this. The honey does not have to be the best-quality closely filtered table honey.

IVY HONEY IN OCTOBER

The following transcript from my diary illustrates just how different one year can be from another; although one has to work to a set of rules, it is necessary to be flexible and deal with any situation as one finds it.

> In beekeeping every year is different and one of the odd things about 1975 has been the unusually dry autumn. Amongst other things, this has produced a record amount of blossom on the ivy so abundant hereabouts, as well as mild weather for the bees to forage on it. As a result there has been a definite honey flow, with a substantial amount stored in brood boxes and empty supers left on for the winter. On some days in October bees were tumbling over themselves to get into the hives and flopping down heavily, as they do on a warm day in June. On 9 October at 11 a.m. all hives were working flat out with torrents of bees whirling up in a funnel from a group of hives, although the shade temperature was only 49°F. Watching one hive closely for a good ten minutes, I noted that at any given moment there were never less than a dozen bees with loads of pollen, as well as twice that number without, scrambling across the alighting board or running down the front of the hive eager to get their booty home. The smell of ivy nectar was very strong for 20 yards down the garden path from the hives, and the noise of the bees flying could be heard from the house a good 25 yards away. The air temperature rose steadily and at 2.30 p.m. reached a maximum of 53°F with activity fully maintained. Never needing much encouragement to forsake gardening for beekeeping, I downed tools, donned a veil and opened nine or ten hives in succession, to find masses of eggs and open brood in each one. One nucleus which I had intended to winter on five frames had eggs and brood on three of them, and not an empty cell anywhere, so I promoted it on the spot to full colony status, and put two drawn frames on each side of the five in a normal brood box (i.e. nine combs). In another stock a young queen was duly marked with a blue spot, having been missed on the 'final' inspection over five weeks before. By 3.30 p.m. the sky had clouded over, the temperature dropped to 51°F and the flow was much less, in fact by 4 p.m. scarcely a bee was flying, although the temperature was still just over 49°F.

That same afternoon I took off a super of clear fresh honey, about one-third sealed, and extracted about 14 lb, which proved to have a specific gravity of 1.38. This is a good deal denser than winter feed syrup (two pounds to a pint of water), which has a specific gravity of only about 1.32. Next morning it was cloudy and stiff, and within 24 hours it had completely granulated to a very white, soft honey. Other supers had this same white granulated honey in the combs, as had the brood chambers, and having set so rapidly the crystals were minute and like soft candy, which no doubt the bees would put to good use next spring. There had been some fanning in the evenings, but obviously nectar so dense as this needs little evaporation, as it is only two or three per cent less dense than ripe honey to begin with. How thoughtful of the ivy, blossoming so late as it does in October and November, to provide nectar with such a high sugar content that it does not need dry evenings and warm nights for evaporation by thousands of bees to ripen it! To my great pleasure I find that the taste for ivy honey grows on one, and instead of being a curiosity it now seems to me to be a honey for the connoisseur. I was at a loss to understand the faint flavour of heather in the background, but checking hive records noted that the ivy honey actually extracted and bottled came from a hive which had been on Dartmoor for the heather flow in August. As a postscript, an entry for noon 20 November spoke of thousands of young bees on nursery flights, as well as occasional bees still carrying in some pollen: 'No shortage of young bees to see stocks through this winter, be it ever so long.'

BEE DRIVING

A hundred years ago, when most people kept bees on natural combs in straw skeps, it was necessary to drive them out at the end of summer in order to harvest the crop of honey and wax. This was done by turning the full skep upside down, pinning a second (empty) skep above it at an angle, with one side hinged to a side of the first, and drumming steadily on the occupied skep with both hands. Aided by a puff or two of smoke, the natural reaction of the bees is not to fly but to run upwards and occupy the empty space above. The honey combs were then cut out with a specially made bent knife, leaving perhaps three or four dark brood combs to be re-occupied when the bees were shaken back an hour later and fed sugar syrup to see them through the winter. As a child I saw this being done by Mrs Crick the carpenter's wife in the village of Brinkley at the end of the First World War, and my father used to buy a honeycomb which would be served up for tea.

The same technique can still be used today, under different circumstances. Beekeepers sometimes take a swarm in late June when they have no time to organize a normal hive with frames and foundation, and two months later the swarm is still living in that same cardboard box. As late as September they may be driven up into a normal brood box with drawn combs and fed for winter, just as in years past, the honey and wax in the box being harvested.

BEES – FINAL PREPARATION FOR WINTER

Feeding

When beekeepers speak of 'winter feeding' they perhaps give the wrong impression. We do not feed *in* winter, but *for* winter, and this should have been completed by the end of September or the first week in October at the latest. If bees are fed too early (in August, for example), the winter food may be used up in unnecessary breeding; if the bees are fed too late (end of October) the syrup will not be evaporated down and ripened. In a mild autumn nature provides winter food from the ivy during the first half of October, and this is extremely valuable both for pollen and honey. However, ivy honey has a higher glucose and lower fructose content than most honeys and so crystallizes rapidly, even in the combs. This is a disadvantage so far as winter food is concerned, but sugar syrup fed during the second half of September or early October will be combined with the ivy honey and produce a much better winter food.

An average stock of bees needs to have at least 35–40 lb of food in the hive about the middle of October, and this is achieved by feeding sugar syrup until the food super (immediately above the brood chamber) has at least six frames solid with food. There will then also be at least as much food again in the actual brood chamber itself. Syrup is made by dissolving white sugar in water at the rate of 2 lb per pint; possible mistakes caused by confusion in converting imperial/metric units can be avoided by pouring granulated sugar into a large plastic bucket up to a marked level and then running in hot water while stirring out the air bubbles until the as yet undissolved sugar/water mixture reaches the same level. It is not necessary to boil, just use hot water from the domestic supply, and stir occasionally (while carrying on with other work); within an hour the sugar will have dissolved.

The important consideration in autumn is that the syrup be taken down rapidly, ripened and sealed over with wax cappings just as honey is in summer. If the hives are close to home (in the garden) then obviously a two-pint round feeder can be used and refilled several times with no great inconvenience, but otherwise the feeder should hold at least one and a half gallons of syrup, containing 12 lb of dissolved sugar. The whole operation should be completed inside a week, and most hives will need the quantity mentioned; incidentally, 12 lb was the official wartime sugar allowance per hive for winter feeding, in the days of rationing.

69 Round 2-pint feeder, with 3-inch eke around to accommodate.

70 Two home-made feeders, made from a coffee tin and a New Zealand lamb's liver container.

Home-made feeders

Perfectly satisfactory feeders may be made from strong, lever-lid tins, e.g. coffee tins, with small perforations in the lid. These are best made by placing the lid (right way up) on a sturdy block of wood and punching 20–30 very small holes with a sharp nail and a hammer, all within a $1\frac{1}{2}$-inch diameter circle in the centre. The holes should not be too large, or syrup will run out; and if the holes are scattered all over the lid, then the slightest tilt of the hive will allow syrup to drip out at the lower level while air bubbles enter at the higher level. A smear of petroleum jelly around the edge of the lid is worth while. In use the tin is filled with syrup, the lid pushed in tight and the feeder inverted over another (empty) tin or bucket to catch the few drops which will run out until the partial vacuum is established, before being placed over the feed hole of the hive crown board. An empty super or box then has to be placed around the feeder to accommodate it before the roof is replaced.

Bought feeders

The small, round feeder already mentioned is suitable for a hive close by, but for general use a Miller-type feeder is best, holding one and a half gallons and shaped like a shallow box, fitting over the whole hive area over the frames and under the crown board. The bees gain access at the front end, where the syrup is deepest, assuming that the hive has the normal slight forward tilt.

Honey or sugar syrup?

Some purists argue that honey is always best for bees but this is not necessarily so, and in mid-winter a pure carbohydrate (like sugar) is metabolized to end products of water vapour and carbon dioxide only, with no protein to leave any solid residue in the bowels at a time and temperature when the bees cannot readily fly to evacuate. No protein is needed in mid-winter. In spring when most of the stored food is used up to rear young bees, protein is needed; honey and pollen are then better than sugar. Thus the best of both worlds is achieved by feeding late in September, when the brood nest is smaller and any incoming food is stored centrally, and on the principle of 'last in first out' is used up first in the November to mid-February period. This leaves the honey and pollen in outer combs for the critical February to April months, when protein food is needed for tissue-building by the next generation.

If a beekeeper should discover at the end of October that a hive has not been fed, or is very light, then it is still possible to feed sugar syrup if a little thymol solution is added, to keep the syrup fresh and prevent fermentation. (See note at end of chapter.)

Storage of supers

Before stacking the honey supers for the winter, check for the presence of pollen in any quantity. Often the central combs in a super immediately above the queen excluder will have several square inches of pollen, and this will go mouldy and become useless by next year. One can scrape it off down to the mid-rib with an old, sharp-edged kitchen spoon, or place any frame with pollen in the middle of a hive food super, just above the brood chamber, before feeding for the winter. The bees will then preserve and later make good use of this pollen.

If there are just two or three honey supers and only one hive in the garden, then they may as well be stored over the crown board (with open feed hole) and under the roof, and on mild days the bees will come up and look after their own empty combs. If several hives are involved, especially at an out-apiary a mile or so away, there is more risk of gale damage, displaced roofs etc., and it is good practice to put the supers wet from the extractor back on the hives for any honey to be taken down, then remove and stack them in piles of eight over wire screen boards (to keep out mice, wasps and wax moths), under a sound roof. This can conveniently be done on a couple of parallel rails placed on concrete blocks to ensure that air circulates under and around and that the boxes are not touching damp earth. Usually a friendly spider will get into the pile from the garden (or can be introduced), and will take care of any insect pests, doing no harm in the process. If you wish to store the supers in a shed or in the house, where the temperature is likely to be higher, then it is important to put a teaspoonful of PDB crystals (paradichlorobenze) on a tin lid in each super to protect against wax moths.

Mouseguards

Over most of Britain there is a 50–50 chance that any hive left with a wide entrance over winter will collect a family of mice by the spring. These will eat away the stored food and build a nest of grass and leaves in the hole they make. Bees then die of starvation or are so demoralized that they dwindle away. Although no mouse would dare to enter in summer when bees are active, they get in with

71 Buckfast feeder, kept permanently on all hives as a crown board. The bees have access via a central hole, up through a wooden block.

impunity during cold weather when the bees are tightly clustered. The standard method of prevention is to put on mouseguards at the end of October. These can be purpose-made strips of zinc with 9-mm ($\frac{3}{8}$-inch) holes, or strips of scrap expanded metal (diamond mesh, used by builders). A better method is explained on page 156.

Queen check

If drones are seen flying as late as the end of September or early October, it is probable that the queen is failing, and the hive should be opened up on a mild day to see if any brood present is normal. There is still time to introduce a young, mated queen, or achieve the same purpose by uniting the doubtful stock with a nucleus or healthy stock considered rather weak for successful overwintering – a late swarm, for example, hived in July or August. The uniting is best done over a single sheet of newspaper. One of the essentials for successful wintering is a vigorous queen, and an old queen may fail to come into lay in early spring, causing the death of the colony before April. Routine checking for the presence of a laying queen is best done in September; if no brood or eggs are present, check again a week after feeding, when the stimulus of an income is likely to bring her back into lay, if she is fertile but just having a rest.

Ventilation

It is generally agreed now that top ventilation is best, so that condensation of water vapour from the winter cluster may be avoided. The bees may have blocked any gauze strip in the crown board with propolis during the honey flow, but will not do so again after September, and the feed hole may be left open under a secure and well-ventilated roof, with its ventilation holes screened by built-in gauze strips. Many beekeepers like to give additional ventilation by inserting four matchsticks between crown board and hive in October.

Secure hives

So far as hives of bees are concerned, the last task of the autumn is to make sure that there is a sound, waterproof roof, weighted down by two or three bricks or tied down by a cord or wire passing right round the hive, against gale damage. Hives in an out-apiary may be at risk from bullocks or sheep, which without intending any harm may use hives as rubbing posts and knock boxes askew, so check fences against farm stock. Woodpecker damage is not a common risk, but it occurs in some areas. It may be prevented by wrapping a length of small mesh wire netting around the sides of the hive. An opened-up plastic agricultural sack will give similar protection.

Equipment

All extracting equipment should have been thoroughly washed out and dried; before putting it away for the winter it is a good idea to wipe metal parts with a cloth dipped in a little medicinal paraffin. Any traces of honey left in crevices or around the edges of lapped metal joints will result in a smelly black deposit next year, as the acid content of honey attacks most metals. Honey stored in sound tin-plated cans, or in the lacquered 28-lb honey tins, will be perfectly all right, but any bare iron surfaces will become black and corroded. Empty hives or boxes can be scraped, wire-brushed and painted with creosote before stacking away for the winter. It is important that the creosote be natural, with no added insecticides.

Conclusion

The foregoing advice assumes that the reader has National or WBC hives, which together make up over 90% of all the hives in use in Britain. With larger hive bodies (Dadant, British Commercial) the actual brood box is considered large enough to hold enough food without an additional food super. In America and Australia it is usual to winter on double Langstroth boxes. However, this only applies to feeding, and the other information still applies generally.

Note

Thymol, referred to earlier, may be obtained from chemists in the form of crystals, which dissolve in alcohol but not in water. A stock solution may be made up by dissolving 20 g of thymol in 100 ml of surgical spirit, and adding 1 ml of this to every three litres of sugar syrup, – about a teaspoon per three gallons. Such a stock solution will keep indefinitely in a well-stoppered bottle.

Finally, now is the time to find out what beekeeping lectures are planned by your local association, or at the nearest technical college, and also, in Britain, to think of some study or reading in preparation for the preliminary and intermediate examination of the BBKA. Your local association secretary will have all the necessary information, copies of syllabus and old questions.

Seven

Autumn – honey and other hive products

Honey is dealt with briefly here, as it is also covered in various aspects in other sections. In this chapter will be found much information on beeswax, pollen, propolis, royal jelly and bee venom not normally included in any one book. Production of some of these items is well within the capability of most beekeepers, and greatly adds to the interest of the craft. As an example of likely yields, I kept a careful check some years ago on exactly what produce was obtained from 16 very ordinary supers of honey, taken off ten hives in Devon at the end of an average year. Apart from 395 lb of honey, there were 9½ lb beeswax, 1½ lb pollen and 6½ oz propolis. The only venom obtained was that delivered personally by individual bees!

HONEY

All honey starts as nectar, which bees collect from flowers, in return distributing pollen to enable fruits to form and seeds to set. Some nectar is also taken from extra-floral nectaries: the openings on laurel leaves, for example. Nectar itself is thin and watery, with the amounts of sugar varying from hour to hour, but low in plum and pear blossom (about 15%), higher in apple, gooseberry and blackberry (about 25–35%), higher still in dandelion, white clover and oil-seed rape (up to 50%) and very high in ivy (can be 75%). To produce honey from nectar, bees have to evaporate off most of the water, convert the sucrose (ordinary sugar) into simple sugars like glucose and fructose, which are more easily digested, and add secretions from their own glands.

A typical sample of well-ripened honey will contain 17–20% water, about 75% fructose and glucose (usually more of the former), 5% other sugars like sucrose and maltose, and up to 3% of the vitally important enzymes, minerals, vitamins, and acid (mostly gluconic), as well as aromatic oils giving the bouquet, flavour and colour. Honies richer in glucose, like ivy and oil-seed rape, granulate rapidly; most honey granulates within three or four months, especially when kept at a temperature around 57°F. Pure heather honey and fuchsia honey never granulate. Enzymes such as invertase, glucose oxidase and diastase are organic catalysts or natural reaction promoters and form a valuable part of honey; unfortunately they are destroyed by heat, within days above 120°F, and within hours above 140°F. Honey kills bacteria and wild yeasts and moulds in three ways: *i* by cell dehydration through osmosis, *ii* by hydrogen peroxide which is steadily formed from glucose oxidase and *iii* by its acidity.

BEESWAX

Mankind has been keeping bees for something like five thousand years, and for almost all this time the most highly prized product of the hive has been beeswax. In the sixteenth century wax made 12*d.* a pound when honey sold for 1½*d.*, and until the second half of the last century the price of wax was usually eight to ten times that of honey. Today it makes only twice or three times as much. In northern countries candles were the only source of artificial light until the introduction of paraffin lamps about 150 years ago, and beeswax candles were easily the best, giving a bright light and a pleasant smell. Poor families had only the light from a log fire, or the smoky flame (and unpleasant smell) of tallow candles made from mutton fat. Thus the Worshipful Company of Wax Chandlers had a virtual monopoly of artificial light, like the Electricity Generating Board of today, and for centuries their business went on unchallenged. But first the paraffin lamp, then gas and finally electricity changed all that. However, wax remains a most important commodity widely used in industry and crafts. Beeswax for comb foundation used by beekeepers absorbs hundreds of tons a year, and other applications range from cosmetics, candles and furniture polish to the wax models in Madame Tussaud's; from batik to cobbler's wax.

Production by bees

Production takes place when young workers are gorged with honey or sugar syrup and hang together in a cluster clinging to each others' legs by their claws, maintaining a temperature of 95–97°F. After about 24 hours the first tiny platelets of wax appear on the wax mirrors situated in four pairs on the under-surface of

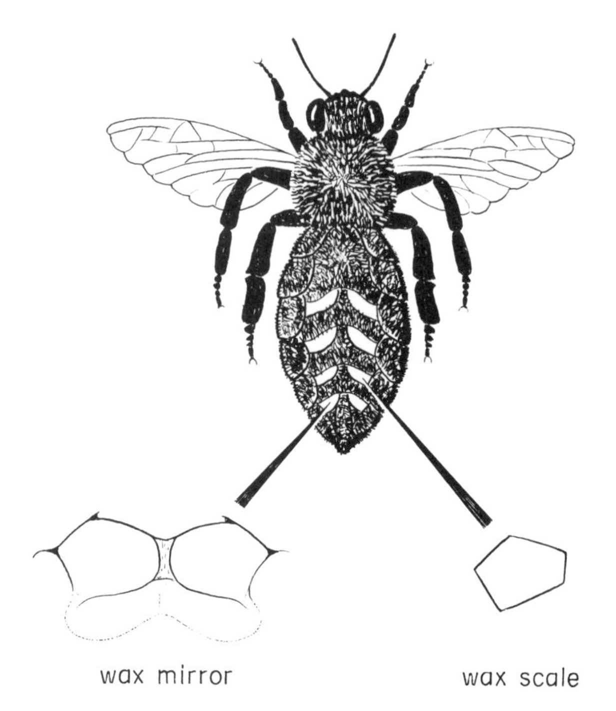

72 Wax-working bee.

the abdomen, and after removal by the worker herself, production is repeated *so long as fresh supplies of honey or syrup are coming in.* This is why it is so important to feed slowly and steadily a nucleus which has to draw out comb from foundation, or a swarm newly hived, when there is only a weak or intermittent honey flow. Although many books give figures varying from 8–20 lb of honey needed to produce a pound of wax, the true figure is only 5½–6½ lb. This was established by Huber nearly 200 years ago and later confirmed by Simmins (1886) and many others. (See Bibliography.)

A worker bee exuding wax scales will normally disengage herself from the cluster and go to the nearest point where wax is required, cling to the work point while impaling a scale on her basitarsal spikes and transfer it via middle legs to mandibles. The scale is gripped so that an edge is presented between the mandibles and can be revolved so that in turn every part can be chewed, saliva and glandular secretions added and the wax 'worked' until it is pliable. If no part-built cells are there, the wax-worker will normally go to the nearest high point, possibly the branch from which a swarm is hanging, before working the wax and depositing it. Each individual wax scale measures only $\frac{1}{8}$ in across (3 mm) and is about $\frac{1}{250}$ in thick (0.1 mm); it takes about half a million wax scales to weigh one pound.

If each wax-working bee produced one batch of eight scales every 12 hours, then it would take 10,000 bees three days to produce one pound of wax (equivalent to seven or eight National deep frames of comb) and to do this they would need to consume at least six pounds of honey. An average swarm of 20,000 bees, weighing four or five pounds, would be carrying not more than two pounds of honey with them at the most, so do not expect too much of a swarm in poor weather. In good weather with a steady honey flow, this same swarm, with 10,000 foragers, 10,000 house bees, and no brood commitment for the first few days, could easily draw out a full box of combs and store a food reserve in the first week. A swarm twice the size could do this and fill a super in the same time. In poor weather the need to feed a swarm is very clear, and even in good weather they might as well use sugar syrup to draw wax.

Wax production at home

This can be increased by carefully collecting every scrap of burr comb scraped from frames, odd pieces of wild comb built down from the crown board and so on, but the bulk of wax produced will usually come from cappings at extraction time, and this may be doubled by good management. It is usually argued that comb-building so diminishes the honey crop as to be uneconomic, and therefore maximum yields are obtained by using supers of drawn comb and making them last as long as possible. This takes no account of the fact that when a colony has a large proportion of young bees, these will often cluster and produce wax anyway, often dropping wax scales on the floor. It is better to go with the bees and let them do what comes naturally – build combs in high summer. One very convenient way of doing this is to use wide-spaced drawn combs in supers by reducing from eleven frames to nine, so that the bees can extend the cells to a greater depth. When uncapping at extracting time, cut deeply down to the woodwork of the frames and harvest the wax as well as the honey. It is usually stated that for every 100 lb honey, about 1–$1\frac{1}{4}$ lb of beeswax may be produced, but this figure can easily be doubled without detriment to the honey crop. Indeed, there is a bonus in that swarming tendencies are lower in stocks which have a normal amount of wax-building to do. At Buckfast Abbey a great deal of beeswax is produced when heather combs are pressed, and the effort required from the bees to draw out some foundation each year is found to be beneficial and has little effect on the yield of honey.

As part of the winter work for improvers it is suggested that a solar wax extractor be made; instructions are given later. This can be used from April to September and when the sun shines will process up to three pounds of wax a day, i.e. as much as any beekeeper with up to fifty hives will have. It is important to appreciate the limitations of a solar extractor, which is very good on odd combs and wax scraps and excellent on dried cappings obtained when extracting. On the other hand, it will not get any worthwhile amount of wax out of old, black brood combs. This is because the wax, on melting, is largely soaked up by the layers of old cocoons and pupa cases embedded in each cell. It is convenient to put such old combs, in their frames, into the solar for half an hour, when the warm comb can easily be cut or even pushed off the wooden frames, which are left clean and ready for the fitting of new wax foundation. A great deal of the weight of these old combs is not wax at all, but layers of old pupa cases with larval faeces trapped at cell base level. Such combs need to be broken up, soaked in cold rain water and boiled to extract the wax. The residues, when broken up, help to make excellent compost. The cakes of wax from a solar wax extractor will usually have some impurities in a thin, dark layer at the base; this can be scraped off and re-melted. Attempts to obtain completely clean wax from a solar are usually defeated by the clogging of any very fine filter screen used, and it is better to re-melt the wax in a double pot (or large jug standing in hot water), and filter it by pouring through a pair of old nylon tights stretched over a wire coathanger bent into an oval shape. For most purposes (e.g. furniture polish), wax straight from the solar can be used, so long as the dark base is cut off.

Wax obtained by boiling old combs is very dark, but may be cleaned and transformed into a more acceptable golden-yellow colour by melting with rainwater (a pint to a pound), and adding to this quantity about two teaspoons (8 ml) of hydrogen peroxide and gently bringing to the boil.

The uses of beeswax

These are so numerous that a complete book could be devoted to this subject alone. In general terms there are five main fields of use, the most important single one being cosmetics, followed closely by church candles, with wax foundation for beekeepers and the pharmaceutical industry next in order. Most of the other applications would come under the heading of 'general industrial use'.

Recipes for furniture polish and cosmetic creams are given in Chapter 8. For a fuller treatment, including candle-making, making wax foundation, wax for showing and the subject of beeswax generally, the reader is referred to the author's *Beeswax*, published by BBNO of Tapping Wall Farm, Burrowbridge, Bridgwater, Somerset.

Care of combs

A pound of wax saved is as good as a pound of wax produced, and a box of good super combs preserved year after year is even better. The great enemy is the wax moth, of which there are two species, the greater and the lesser; in Britain the lesser is usually the greater danger, and it is found almost everywhere. Its appearance closely resembles that of its cousin, the common clothes moth, being about half an inch in length and usually of drab greyish-brown colour with a silvery sheen. When disturbed it runs rapidly about on floor board or comb surface, often with a deceptive zigzag avoiding action that makes it surprisingly difficult to catch. It is very flat in profile and able to squeeze through narrow cracks. In the larval stage, these pests tunnel through wax combs, usually just to one side of the mid-rib, eating some wax but spoiling much more. They protect themselves with tunnels of spun fibre reinforced by their excrement, making it very difficult for bees to get at them to throw them out. In extreme cases a box of combs can be reduced to a foul tangle of criss-crossed web tunnels with literally a thousand or more pupae lying flat on the frames or sides of the box, encased in tough cocoons. A surprising feature is that the larva can eat into the actual wood of the frames, making a shallow, tapering oval depression which scars the woodwork permanently. Techniques for preserving wax combs are given in the 'Autumn work' section on page 123, but in general terms cold is the enemy of wax moths, which thrive best in a warm place. To wrap combs or whole supers in old newspapers and store in a spare bedroom is a recipe for disaster, as eggs or larvae will almost certainly be present, although not noticed. This point is made because beginners have sometimes gained the impression from reading bee books that there is a protective virtue in newsprint. There isn't. It merely excludes other moths, and provides a cosy environment free from spiders in which existing eggs and larvae can develop and the mature moths can mate and reproduce a hundredfold until the combs have been virtually destroyed.

Apart from protection from wax moths, it is important to remember that wax is very brittle in cold weather and needs gentle handling. In summer conditions, or after a few hours in a warm room, the warmed wax is more pliable and much less at risk. Fixing sheets of new wax foundation into frames should always be done in a warm room, or on a warm day outside.

As mentioned in the section on pollen, any stored outside the actual brood nest is likely to be hard and unusable next year, and often covered with grey mould. The answer is to scrape down to the wax mid-rib and remove any pollen stored in super combs, so that the bees will readily draw out new cells in early summer. With old brood combs this can be done only with great difficulty. Another method is to freeze the combs (outside in winter, or overnight in a deep-freeze) to destroy wax moth and then inoculate with a teaspoon of pollen-mite debris (a fine yellow powder) obtained from another beekeeper, or sometimes found present in any case, as the minute pollen mites seem to be able to extract some food from pollen. These mites work well at low temperatures, e.g. in a corner of the garden shed, and clean out pollen-clogged old combs very well, leaving a layer of very fine pollen dust on the floor.

Sterilization of combs with acetic acid will also have the effect of killing any moth larvae and other pests.

POLLEN

What is pollen?

Pollen is the male sex cell of plants and trees, produced as minute loose grains on the anthers of stamens. These apparently lifeless grains are actually living cells, each containing two nuclei in a mass of cytoplasm, protected by a thin semi-permeable membrane called the 'intine', inside a tough protective shell (the exine) which carries the characteristic pattern of pores or furrows. On a receptive surface (the sticky stigma of a compatible flower) the cell germinates and the two nuclei are carried inwards via a pollen tube which grows out of a pore into the stigma, rather like a miniature root from a pea or bean. When the tube reaches the ovary, the male nuclei fuse with the female nuclei in the flower's ovary, leading to the development of seed and fruit. The size of pollen grains can vary enormously, from 3 microns for forget-me-not to 150 microns for hollyhock and marrow. One micron, written 'μ' is $\frac{1}{1000}$ mm or $\frac{1}{25000}$ in.

Composition of pollen

The cytoplasm or interior mass of pollen grains is a very valuable food indeed. It contains:

Protein

Up to 40% in the form of at least 21 identified amino acids such as arginine, histamine, leucine, valine, etc., which are essential for normal growth and development.

Vitamins

Instead of quoting each vitamin, it is easier to state that pollen contains every known vitamin, except that the B_{12} content is very low. Vitamin C is present in large amounts, and probably accounts for the slightly acid flavour.

Minerals

These account for 3–8% of the weight, and include potassium, phosphorus, calcium, magnesium, manganese, iron and over 20 trace elements.

Fats

Some pollens, dandelion for example, contain a considerable amount of fats; these are valuable but not essential to bees, as bees can synthesize body fats very easily from the sugars in honey.

Use made by bees

Bees collect pollen as a protein food and major source of minerals and vitamins, and in the process pollinate the plant or tree visited. It is known that the economic value to the community of pollination is several times greater than that of the honey collected and extracted for use. The biggest single use of pollen is in the production of brood food, on which larvae are fed; young nurse bees (three to six days old in spring, six to nine days old in high summer) eat 10 mg a day and produce via their hypopharyngeal glands a rich creamy liquid containing 4 mg of protein per day. It takes about 100 mg of pollen to raise one worker bee, i.e. 1 kilogram of pollen to raise 10,000 bees, or roughly 1 lb to raise 4,500. Thus a normal colony needs upwards of 40 lb of pollen a year just for this purpose. In addition, pollen is eaten later in the year to build up the protein food reserves of overwintering bees and to provide royal jelly (brood food) for the queen throughout her life. Not all pollens have the same nutritive value, the best pollen coming from fruit trees, willow and clover, followed closely by dandelion and elm. That from alder, hazelnut and pine trees is not so good.

The nutritional value of pollen to bees is known to decrease fairly rapidly. Haydak (*Hive Products*, Apimondia 1961) found that after one year pollen retained only 25% and after two years less than 5% of its original value, so far as the resultant secretion of brood food was concerned. Subsequent work has shown that two amino acids (L-lysine and L-arginine) deteriorate rapidly, and the addition of these substances to pollen not more than 2½ years old can restore most of its value. So far as practical beekeeping is concerned, this point emphasizes the value in spring of pollen brought in by the bees in late autumn (October) from ivy and Michaelmas daisies.

Uses to humans

Pollen contains five or six times the amino-acid content of lean beef, but it is as a broad spectrum bio-stimulant that pollen is most highly regarded, even when taken as a food supplement in small quantities which would not in itself provide anything like the total amounts needed. For example, patients recovering from surgery have shown significantly greater increase in weight and recovery of strength generally, when given regular but small doses of pollen in their diet. Reports from Holland and Sweden speak of the value of pollen in the treatment of prostatitis, impotence and sexual conditions which other treatment had failed to help; patients experienced a return to sexual normality, with elimination of pain and sensitivity. Throughout clinical tests, the men were treated with pollen extract (three pills a day) and no side-effects were noted during several years. Pollen has also been effective in the treatment of chronic alcoholism, a condition notorious for its accompanying lack of vitamins and amino acids. The difficult withdrawal or abstinence syndrome, normally lasting about a week, is very much reduced by the intake of two or three grams of pollen per day for this period. The discovery of gonadotropic hormones in the pollen of date palm trees (Ridi, 1960) gives a scientific basis to its ancient use by the Bedouins in treating male sterility.

Hive Products, a publication from Apimondia, the world bee authority with headquarters in Romania, recommends one teaspoon of beebread mixed with honey and taken four times a day for anaemia, colitis and gastritis. Many sufferers from hay fever find that great relief is obtained by eating half a teaspoon a day of a pollen and honey mixture from February onwards, and it is convenient to mix equal quantities of warm, run honey with pollen scrapings (stored in the deep-freeze since extracting time the previous year) and supply it as 'natural pollen preserved in honey'. Obviously it would be unethical to make any claim for this product as a remedy for hay fever, but it is a good natural protein food in its own right anyway. My customers include a well-known TV personality as well as a senior teacher who used to teach in June with a handkerchief in one hand and a piece of chalk in the other; now he manages just with the chalk! Some honey creams contain 10–30% crushed pollen as a skin food, smoothly blended in. Pollen tablets on sale at chemists contain $\frac{1}{4}$–$\frac{1}{2}$ g of pollen plus sugar, gelatine or gum arabic and other matter to give a firm texture and smooth coating.

Pollen collection

The standard method is by means of a pollen trap placed at the hive entrance, whereby pollen loads are scraped off bees' legs as they enter and fall through a grille into a tray. At times when pollen is plentiful, enough will get through the trap to satisfy the needs of the hive, and yield a surplus for the beekeeper of up to 60 or 80 g a day in early summer. The best pollen-collecting stocks are usually headed by young queens.

The main purpose of collecting pollen at a time of surplus is to have it available to feed back to the bees at a time of shortage (early spring, in some places), when pollen shortage may be a factor limiting the expansion of the colony. More efficient use of it may be made by mixing it with 'artificial pollen' to make this more attractive to the bees. Another reason for collecting pollen, of course, is to have it for sale or personal use.

There is a distinction between pollen scraped from bees' legs and collected before storage by them, and pollen actually stored by the bees in combs and extracted from them. After storage in cells it is technically known as 'beebread', and has slightly different properties, containing some added honey to preserve it, glandular secretions from young bees and possibly also some propolis. Medical workers have reported beebread to be a more effective bio-stimulant than dried pollen: one report also mentions that beebread contains an anti-anaemia factor not present in dry pollen. It is not suggested that pollen should ever be taken from combs which have been bred in by the bees, but in single brood chamber management pollen is often stored by the bees in the central combs of the first super immediately above the centre of the brood nest. When combs are centrifuged, pollen remains in the cells, and if left in a honey super after extraction will inevitably go mouldy and provide a great deal of unnecessary hard work for the bees which have to remove it next summer, so that it may as well be removed and put to good use. After honey has been extracted, the pollen may be scraped off with an old, sharp-edged kitchen spoon, leaving the wax mid-rib foundation to be drawn out again as fresh comb next year. This pollen, plus the thin side walls of wax that come with it, should be stored in large plastic boxes in a deep-freeze, where it will remain in good condition until needed. Such combs always have some pollen still left, and together with combs of scattered pollen cells not worth

73 A pollen trap placed under the hive provides the bees with only one entry – through a 5-mm mesh screen which brushes the pollen from their legs and drops it into a collection tray. (Australian Information Service photograph by Mike Brown: Agriculture – miscellaneous 18/5/77/5.)

scraping are best placed in the centre of a super immediately over the queen-excluder, as a reserve food chamber before autumn feeding. The bees then remove any pollen fragments from the midrib and cover open cells of pollen with ripened syrup before capping.

General points

Nectar collected naturally by bees normally contains some pollen, but up to 50% of this is removed within 15 minutes, i.e. during the flight home. It is removed by the proventricular valves between stomach and honey sac, and the pollen is digested by the bee. There is ample pollen left, however, to serve as a basis for identification of honey sources, using microscopic techniques. The extreme durability of pollen grains, the exine of which survives treatment with caustic soda and is non-biodegradable, has been made use of in archaeological techniques of pollen-dating. Pollen known to be from a certain epoch can be used to date strata in excavations.

Sometimes pollen in stored combs will be dealt with by tiny pollen mites, which apparently extract some food from them and leave an extremely fine powder, which can be tapped out of the cells; this does no harm and is much easier than having to deal with the pollen stones left after mould has got to work.

PROPOLIS

Pliny the Elder, in his *Natural History*, written nearly 2,000 years ago, described propolis as a horrible nuisance. Even in recent times it was regarded by most beekeepers as something sticky which was scraped off with curses and thrown away; then in the mid-1970s it quite suddenly became a valuable by-product of the hive, as two or three different agencies offered up to £1.50 an ounce and exported it to Scandinavian countries, where it was understood to be used as a natural antibiotic in the treatment of respiratory infections. Within a couple of years the demand slackened off, and for several years now few have bothered to collect it, yet bee literature from Central Europe, Russia and other Eastern European countries continues to emphasize its importance, and international conferences have been held in Bucharest and elsewhere, and scientific papers presented dealing with its use in medicine, its chemical composition and research into its many properties.

74 Propolis being scraped off a frame by the author. (Colin Rome photography.)

Background facts

For the benefit of new beekeepers it should be stated that bees collect propolis (bee glue) from the bark and buds of trees and use it in their hives to block up small holes and crevices, fasten down frames and cover boards, and sometimes to restrict hive entrances by building a wall with a just a few pop-holes for traffic. The famous blind Swiss beekeeper Huber in 1798 described the substance as a reddish, unctuous, odoriferous varnish, applied to strengthen the mouths of cells. When he, or rather his assistant Burens, failed to observe bees gathering propolis from trees, he had twigs of fat, red poplar buds cut and placed in vases in front of his hives, and then described how the bees pulled the sticky material from the buds and packed it in their pollen baskets in small glistening globules 'the colour and lustre of a garnet'. He also observed bees mixing fragments of old wax with propolis, when working it in the hive. He noted that propolis partly dissolved in turpentine, ether and spirit, giving a golden colour to the solution. He also noted that fresh wax scales (removed from workers' wax glands, or rather the external 'wax mirrors' on which they are formed), dissolved completely in turpentine, yet even the whitest fragment of worked wax taken from the newest piece of comb, crumbled and left a residue. The name 'propolis' was derived centuries ago from the Greek *pro*, 'in front of', and *polis* – 'town', or 'city', indicating this use of the substance against other insects, and the weather.

Basically propolis is a resinous substance

with a pleasant aromatic smell, in colour varying from light brown to a dark chestnut red. In general terms it is composed of 55% balms and resins, 30% waxes, 10% ethereal oils and 5% pollen. It is heavier than water and water-repellant. Although very sticky at hive temperatures, it sets quite hard below about 15°C (59°F) and is brittle below about 5°C (39°F). It is carried in globules on the pollen baskets of bees, usually in very warm weather. Sometimes bees will collect a tar-like substance from hot macadam, asphalt and even some paints. Presumably this is then used only as a building material, yet the organic origin of these products millions of years ago may mean that the bees can still extract from them the substances needed.

Sources

In the USSR most propolis is collected from birch trees, the actual figures being: birch 65%; poplar 15%; birch/poplar mixed 15%; other sources 5%. In Western Europe the major source is reckoned to be poplar trees, but horse chestnut, alder, birch, plum, cherry, willow, spruce, elm, oak and conifers all contribute. Flavinoids extracted from sticky buds have been shown to have antibiotic and antifungal properties: the advantages to the plant are obvious, and would constitute a survival factor likely to develop over tens of thousands of years. Although sticky buds produce most, bees also collect gummy exudations from the bark of, for example, plum and cherry trees. Propolis is recycled both in and out of the hive. On hive walls and frames there are often small round blobs of propolis, not having any immediate function but obviously stored for convenience and future use. On a hot day in July it can be both instructive and amusing to sit down on the lawn beside an old crown board, with a magnifying glass, and watch bees tearing off loads of propolis to take back to the hives. In the sun it softens and becomes tacky, like warm carpenter's glue, and a bee will tug away with its mandibles, pulling like a small tug-o'-war athlete with braced legs, and then suddenly a small sticky lump breaks away and the bee falls over backwards with complete loss of dignity. The sticky fragment is then transferred by the front legs to middle legs to back legs in movements so swift that one has to see them repeated several times to master the exact sequence of movements. Finally it is packed on the pollen baskets as one or more round drops, glistening like miniature brown dew drops. They are usually content with loads much smaller than pollen loads, and need help at the hive to unload. If it is quickly and eagerly unloaded by other worker bees, the propolis collector goes out for more, but if kept hanging about for some time, the bee is discouraged and stays at home, or goes out for nectar and pollen instead. It is probable that only a few bees at a time collect propolis, on very warm days, and the intake is subject to supply and demand, like that of pollen.

Another source of propolis is thought to be the balsams contained in the husks or exine (outer wall) of pollen grains, released when they are digested in large quantities by young workers manufacturing the protein-rich brood food for the larvae; young nurse bees need about 10 milligrams of pollen a day for this purpose. One authority states that oily droplets of pollen balsam are pushed back into the honey sac by the proventricular valves, regurgitated and mixed with bee saliva containing a digestive ferment before being deposited to harden in round blobs on the hive wall.

Use made by bees

Within the hive, propolis is used to block up small holes and cracks. Any opening less than about 4.5–4.8 mm is usually blocked with propolis alone, but small quantities of added wax are sometimes found in apertures above 3.5 mm. In the range 4.8 mm to about 10 mm, any opening is usually left clear for bee traffic, but may have small pieces of either wax or propolis deposited in it. A gap above 10 mm will normally be filled with burr comb made of wax, and the larger the space the more likely it is that well-formed cells will be built. An exception is made at actual entrances, or just inside the hive, where some colonies, especially of Caucasian bees, will sometimes build a complete curtain or wall of propolis measuring up to two centimetres in width and several centimetres long, to keep out either pests or unwanted draughts. In Central Africa my Adansonii bees used propolis extensively at hive entrances, more to keep out beetles and other insects than to control ventilation, in my opinion.

When an intruder too large for the bees to move has died or been killed inside the hive, the bees will encase the body in propolis, so that the flesh does not rot and cause an obnoxious smell but slowly mummifies. A mixture of propolis and beeswax has a higher melting point and more strength than either product by itself, and on one occasion in the tropics I found natural comb fastened by this mixture inside a black metal box in full sun, obviously reaching a very high temperature at midday.

Apart from use as a glue to secure loose parts, or as a domestic 'filling compound', the really important use of propolis is as an antiseptic varnish. The inside walls of a new wooden hive (or of a natural cavity in a tree or building) are thoroughly varnished as soon as the colony grows large enough to occupy most of it, and even a small colony on four or five frames will varnish its immediate area. The inside walls of every wax cell are varnished with a very thin layer of propolis within hours of being built, before the queen can lay in them, or any nectar can be stored. This can be illustrated by a very simple experiment. Take a sample of apparently pure white comb, built in the last few hours in high summer, and melt on a watch-glass over a spirit lamp: there will be a visible residue remaining in the molten wax. As a control experiment, repeat this with some wax scales dropped by wax-working bees on a paper sheet inserted in June on a hive floor (especially under a recently hived swarm). These unworked scales, as exuded from wax glands of young worker bees, will readily melt into a colourless liquid with no residue. The main purpose of varnishing the hive and cells with propolis is thought to be internal hygiene rather than just strengthening or waterproofing. It is known that propolis contains natural antibiotics, and its use in a hive represses the bacterial and fungal growths otherwise to be expected in a warm and humid environment crowded with thousands of living creatures. The natural resistance of some colonies to infectious brood diseases may be at least partially due to propolis or propolis extracts incorporated into the wax of brood comb. *Mellissa coccus pluton*, the causative agent of EFB, has been shown to be sensitive to propolis, but so far no such effect has been demonstrated on AFB, even when fed in dilute honey to a colony of bees. The great French authority Professor Rémy Chauvin carried out a series of experiments to monitor the presence of bacteria and fungal flora on various insects and discovered that honeybees are unique in this respect, being virtually free of such organisms.

Use to mankind

In ancient times, propolis was used in medicine and surgery as an antiseptic and for its healing properties generally. It is thought that skull-trepanning operations, incredibly performed in prehistoric times, were possible only with the antiseptic aid of natural propolis from wild bee nests. When knights of old wounded in battle were said to be cured with sweet-smelling balsams, perhaps propolis was the active principle.

Violins made by the great Stradivarius were varnished with propolis harvested by bees in the Cremona region of Italy. During the Anglo-Boer War effective use was made of an antiseptic dressing prepared from petroleum jelly and propolis; subsequently the medical profession tended to despise what is termed 'folk medicine'. However, it has enjoyed a revival, and today a great deal of medical use of propolis is being made once again, particularly in Eastern European countries.

At first the main uses of propolis ointment were on horses and cattle, in the treatment of infected wounds, hoof fissures and injuries generally. Then the unexpected success of these measures, and possibly also the relative economy and wide availability of propolis, led to extensions into the field of human medicine. Currently propolis and its derivatives are being widely used in a number of countries, and the success obtained cannot be fully explained by analysis of the natural product itself; empirical medicine is running ahead of theory. Here are a few specific examples of the medicinal uses of propolis from reports which, backed up by details of carefully planned experiments with control groups, have been published in Bucharest by Apimondia, from the proceedings of international conventions held in Bratislava (1972), Madrid (1974) and Bucharest (1976):

a An alcoholic solution of propolis used twice a day as a spray against bedsores achieved remarkable success.

b Propolis taken internally has also proved to be a useful stimulant in the development of immune reactions over a wide field.

c A mixture of 20% propolis to 80% ethanol agitated at room temperature for 48 hours and filtered, was successfully used for the treatment of colitis. Thirty drops were taken in a glass of milk or warm water three times a day, after an initial single dose two days previously as a reaction check. Of 45 patients so treated, the effect was rated 'very good' with 25, 'good' with 12 and 'no effect' with 8. (Nikolov, Bulgaria)

d 100 g petroleum jelly melted with 10 g propolis, heated to 70–80°C, stirred and filtered through a fine sieve, and cooled to an ointment for external use, proved successful with ulcers, lesions and burns.

e A general tonic made from malt extract, calcium triphosphate, propolis and honey was found to be very valuable, more effective than other standard tonics. (Several contributors)

f A persistent sore throat with concurrent high temperature was cured in six hours by gargling with an infusion of crushed propolis and warm water – a yellow liquid like tea. A similar infusion caused painful smarting, but successfully cured an eye infection of long standing. (K. Lund Aagaard, Denmark)

g The sensation of numbness in the mouth after sucking a small piece of propolis is well known, and this anaesthetic property can be made use of, a 1% solution in a water/alcohol mixture having much the same effect as procaine when injected sub(Common practice in USSR)

h Many observers have noted that propolis added to food and taken regularly has a very beneficial effect on urinary tract infections, specifically on prostatitis.

General points

The 55% content of resins and balsams in propolis is composed of varying amounts of many complex organic chemicals, but most are derivatives of flavins, vanillins, chrysin and allied compounds, with aromatic, unsaturated acids like caffeic acid. More recently, ferulic acid has been noted. The steady application of modern techniques of gas chromotography is revealing the presence of more and more complex compounds, and there would be little point in cataloguing all the constituents so far identified. Quite apart from the resinous and other raw materials involved, salivary secretions and secretions from the worker bee mandibular glands are added in the hive, and more wax often incorporated. Wax contained in propolis itself, as opposed to wax added later by the bees, is not identical with normal beeswax and has a different melting point, iodine value, etc. Presumably this wax is derived from digested pollen and does not come from the actual wax glands of the bee.

Propolis is reasonably soluble in alcohol, acetone, ether and turpentine. It is broken up by alkalis, so that a propolis stain can be removed by household ammonia or washing soda. Propolis on metal ends is best removed by boiling in a saucepan of water with a small handful of washing soda crystals dissolved in it.

Propolis has been found to reduce the amount of queen-cell building, and also of laying-worker development. I have also noted that queenless stocks on old combs with a high propolis content give rise to laying workers less readily than stocks on newly drawn combs, but this could be due to the effect of queen substance retained on old combs much walked on by queens.

BEE VENOM

Literally everyone knows two things about bees – that they produce honey and that they can sting. About one person in five knows that the sting is barbed and that the bee stings only once and then dies, whereas a wasp sting is not barbed and can be used over and over again. The information which follows brings the story up to date.

Fresh bee venom is a slightly yellowish liquid with a specific gravity of 1.1313, containing about 44% of dry matter when evaporated, which it does rapidly in air. It is soluble in water but not in alcohol, and has a bitter taste and strong odour. Each bee 'load' is about 50 micrograms ($\frac{1}{20}$ mg or $\frac{1}{600,000}$ oz). Venom is produced by bees over three days old and under three weeks old, and a protein diet (of pollen) is essential to this production. Older bees retain venom in their venom sacs (and can use it) but the actual venom gland atrophies.

Venom is produced in the so-called 'acid-gland', a coiled milky-white tube lined with modified epithelial cells, leading via a thin transparent tube to the enlarged sac (looking like a plastic balloon) in which venom is stored. From this sac leads a funnel-shaped duct, chitinized where it enters the actual bulb of the sting. This bulb is hard and not easily compressed (see Fig. 75). There is no mechanism for causing the sac to contract and thus squeeze out venom; it is just a storage tank with walls which collapse when venom is pumped out by the reciprocating mechanism of the sting itself. Thus there is no danger of squeezing out venom when taking a sting out of one's flesh, as many books suggest, and the all-important point is to *get it out at once*, before the automatic pump, which carries on by itself even when the bee has torn itself away, can pump in its full load – about two minutes.

Venom has a most complicated composition, known in any detail only in recent years with the advent of sophisticated techniques such as electrophoresis, gas chromatography and mass spectography. So far at least 26 molecular weight fractions have been identified, and these fall into five main categories, each with a different purpose.

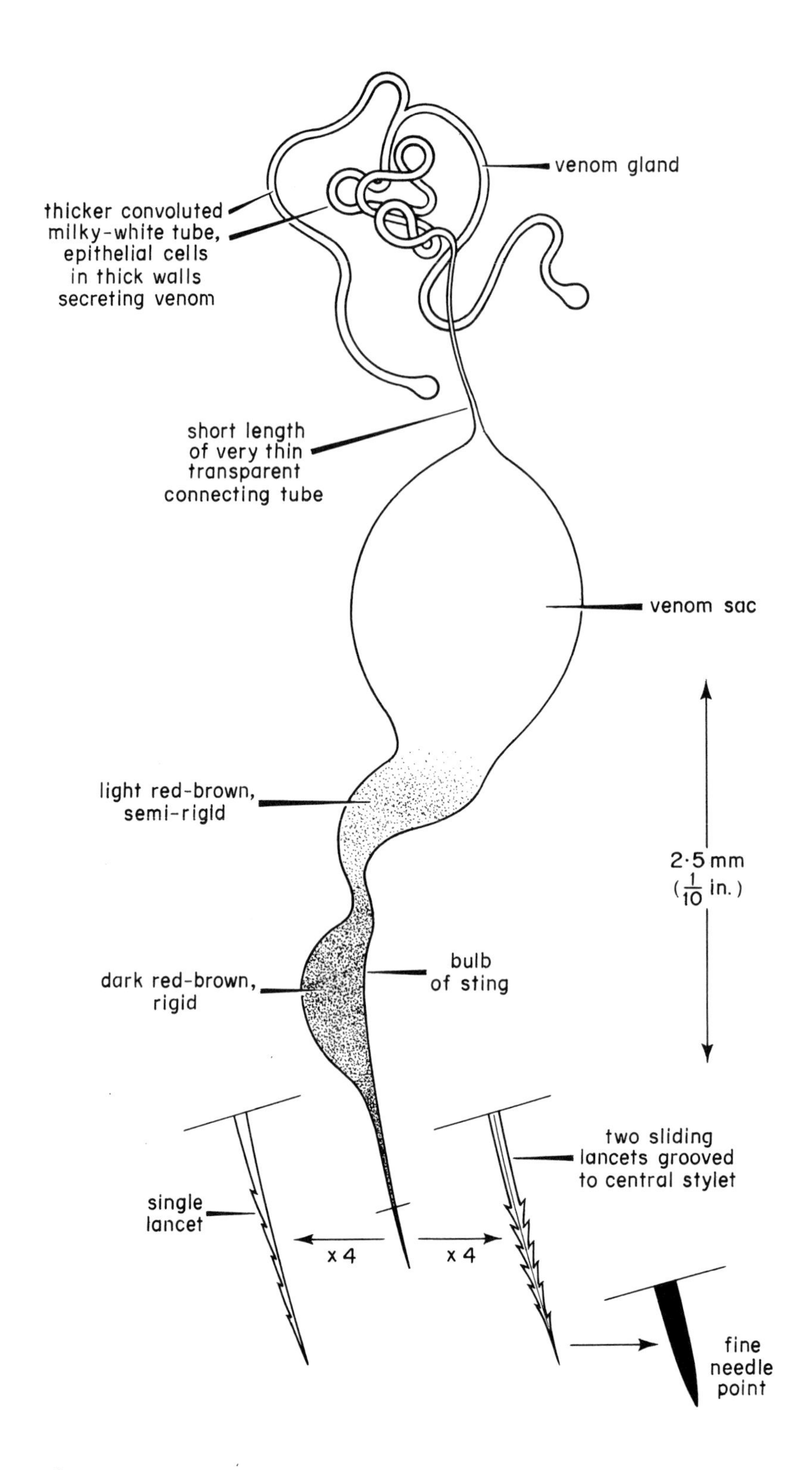

75 Bee sting and venom sac. Note the fragile, transparent venom sac, in appearance rather like a plastic balloon.

a *Directly toxic.* Low molecular-weight polypeptide chains (amino acid radicles).

i *Mellitin* (which accounts for almost 50% of the total dry weight), a neuro-muscular blocking poison.

ii *Apamine* (2%) and *dopamine* (1%) – both neuro-toxic compounds.

iii *Histamine* and *novadrenaline* (traces).

Collectively these substances are responsible for the initial pain, redness and slight local swelling, even when a beekeeper has gained immunity. Against very small creatures (other insects, for example) they are lethal. Against larger creatures it is a question of body weight, and the killing dose by immediate, direct action is about 1 mg per kg of body weight. What is instantly fatal to another bee (weighing $\frac{1}{10}$ gram) and crippling to a mouse, is only painful to a dog or human. So other effects are necessary to defend the colony of bees against larger creatures.

b *Spreading agents – enzymes.* These are proteins of high molecular weight which break down the cell structure and enable the poison to spread.

i *Hyaluronidase* (3% dry weight) degrades and breaks down the polymerized hyaluronic acid which forms the barrier tissue in the cellular matrix, enabling the venom to spread, but only locally.

ii *Phospholipase A* (12% dry weight) catalyses the degradation of phospholipids (fats), also breaking down cell membranes and enabling toxins to spread.

iii *Acid phosphatases* having similar effects.

Collectively these make the toxins in venom more effective, but fortunately most of us have body defence-mechanisms which protect against these substances, which in any case are not concentrated enough to affect larger mammals (bears, homans). For this reason yet another effect is called into play, viz. *c*.

c *Protein irritants* (30% dry weight) which can set up allergic reactions causing the body to poison itself by massive release of histamine from its own cells. This can be effective against large mammals, and cause collapse or even death in extreme cases.

d *Steroids* (cortisones and related compounds), which have less dramatic effects, as yet imperfectly understood.

e *Alarm pheromone* – a chemical called isopentyl acetate with a strong odour similar to that of cellulose 'thinners' or some brands of nail varnish (which contain a related compound, amyl acetate). This is the bugle call, bringing in other bees to sting, just as in aerial warfare a specialized single aircraft with an expert navigator may mark the target for the rest of the squadron to attack.

The allergic reaction, *c*, is the most dangerous, and worth going into some detail, as it is this which is responsible for the serious reactions which affect a very small percentage of human beings. Our bodies are programmed to reject alien proteins as a defence mechanism; this is the great difficulty in operations such as kidney- and heart-transplants. In our blood we have about 70 g per litre of proteins, roughly two-thirds albumin and one-third globulin, and in this connection we are concerned with two globulins: *immuno-globulin G* (or IgG), which is normally involved in protection against bacterial infection; and *immuno-globulin E* (or IgE), normally protecting us against parasitic protozoa.

Such substances as these are called 'antibodies', and the whole basis of injections against typhoid, tetanus etc. is to build up the appropriate 'antibody' as a defence against any subsequent infection. When foreign substances such as bacteria or proteins enter the blood, the antibodies attack and neutralize them, while at the same time the reaction triggers off the mass production of further antibodies. Our genetic make-up and case history determines whether we produce more IgG or IgE. In general terms, asthma and hay fever sufferers tend to produce more IgE and other people more IgG. IgE attaches itself to mast cells (lung tissue, skin), whereas IgG is mobile.

As already stated, most of us produce much more of the desired IgG and successive stings build up our resistance so that although we may feel slight initial pain, after a few minutes we may not even remember where we were stung. Our immunity even lasts over winter, or at least builds up again rapidly after the first sting of spring.

The unfortunate few who produce more IgE suffer progressively worse reactions after successive stings, leading, in extreme cases, to collapse. This is caused by the interaction of venom protein molecules and IgE antibodies attached to their cells, whereby these cells are broken down and release histamine on a massive scale.

Antidotes

Any anti-histamine or decongestant treatment helps, for example two aspirin tablets or one Triludan tablet, taken an hour before going to the apiary. Alternatively, an adrenalin medihaler (obtainable on the National Health) is effective immediately, especially against congestion of the lungs. A proprietary spray, such as 'Waspeze', gives local relief. If taken to hospital as an emergency case, probable treatment would be an injection of adrenalin. Folk remedies, such as rubbing with an onion or treating with sodium bicarbonate, are unlikely to have more than a psychological effect, but this is not to be totally disregarded, if nothing better is available.

Long-term treatment

Since 1980 it has been possible to give immuno-therapy in which minute doses of pure bee venom are injected subcutaneously at regular intervals, the dose being gradually increased until the patient can tolerate 100 micrograms (equivalent to two full stings). Usually it is possible to start with an injection of $\frac{1}{10}$ microgram ($\frac{1}{500}$ of a natural sting), and increase gradually to 1 microgram, then 10 micrograms, and so on. Under this regimen the amount of IgG is steadily built up until the body regularly manufactures enough to cope with stings, so that no beekeeper today needs to give up because of an allergy.

(H. R. C. Riches, MD, FRCP, lectures on 'Recent Advances in the Treatment of Bee Venom Hypersensitivity', to Central Association of Beekeepers, London, 27 March 1981, and at the Seale Hayne Agricultural College, Devon, July 1984.)

Venom production

Bee venom is obviously needed for the treatment of sting allergy, but much work is being done (mostly outside Britain) on the use of venom to relieve rheumatism and other conditions. Among beekeepers it is said that rheumatism seldom afflicts people regularly stung by bees.

One method of collecting bee venom is to tie a large sheet of thin rubber over the open top of a strong hive, and rub this gently with a nylon cloth. This charges the rubber electro-statically and annoys the bees. Hundreds of them will immediately sting the rubber sheet, the actual lancets penetrating the rubber so that droplets of venom appear all over the upper surface. The stings are not trapped by their barbs (as in flesh) and can be withdrawn so that no bee is killed. Within minutes the venom evaporates, leaving minute scales of dry matter, which flake off when the rubber sheet is removed and folded. Not many beekeepers will wish (or have the skill) to produce and market bee venom, but the price per gram is higher than for any other hive product.

ROYAL JELLY

It has been known for a long time that both worker bees and queens can develop from fertilized eggs, and that the determining factor is the diet given during the larval stage. For their first two and a half days, larvae lie in a pool of brood food, a small pool in the case of worker cells and a much deeper pool in the case of queen cells. By the third day the queen larvae are floating on an increasing amount of this food, called 'royal jelly', while workers no longer have a food reservoir alongside, though they are fed at regular intervals: queen larvae are fed about ten times more frequently than worker larvae, which receive only the 20–40% white component of the brood food for the first two days. This white component is a mixture of secretions from both hypopharyngeal and mandibular glands, as opposed to the colourless component which is a mixture of secretion from the former gland only with nectar regurgitated from a nurse bee's honey stomach. The secretion from the hypopharyngeal glands is mostly protein; that from the mandibular gland contains fats. Regurgitated nectar will usually contain enzymes, and some pollen grains. Queen larvae have their cells sealed after the fifth day, but go on eating and gaining weight for at least a further 12 hours.

The precise difference between the brood food fed to worker and royal larvae has not been finally determined, but recent work by J. Beetsma (*Bee World* 60(1): 24–9, 1979) indicates that queen/worker differentiation is controlled by the varying activity of small glands (*Corpora allata*) situated near the proximal end of the oesophagus. These glands secrete a substance called juvenile hormone, triggered off by sugars present in food. Higher levels of juvenile hormone secretion produce queens, and much lower levels produce workers. Thus it is now considered that sugar intake over the first two or three days plays a major part in making young larvae develop into royal larvae and subsequently become queens. However, it is known that royal jelly given to a $2\frac{1}{2}$-day-old larva in a queen cell has a much higher content of three substances, viz: pantothenic acid, biopterin and neopterin, than brood food given to larvae of the same age in worker cells. It is thought that these substances (and possibly others yet to be identified), play their part in full queen development over the third to sixth days. After emergence from the royal cell, a virgin is quite capable of helping herself to food in the hive, or if caged by herself, to candy in the cage, but not of foraging for food. However, she will normally be fed by the young bees before and after mating, again on this protein-rich, creamy food called royal jelly.

What is royal jelly?

A simple analysis shows that it is a creamy, milky white liquid with a faintly pungent odour and a fruity taste, probably because of its acid content. Apart from being two-thirds water, it contains $12\frac{1}{2}$% protein and about the same amount of sugars, $5\frac{1}{2}$% fat and 1% minerals. The remaining factors include: 10-hydroxydecanoic acid (amongst other things an antibiotic), possibly derived from the mandibular glands; gamma globulin, which is known to confer increased resistance to virus and bacteria; nucleic acids like DNA and RNA; cholesterol; pantothenic acid; biopterin; neopterin; and a number of other substances known only as peaks on chromatograms and not yet positively identified. Royal jelly is also rich in vitamins, especially B and C, though lacking in E.

Royal jelly production

Small quantities can be obtained from time to time during summer from swarm cells cut out in normal management, but for quantity production it is necessary to induce the production of large numbers of queen cells by young bees well fed on pollen and honey (or sugar syrup). Two simple ways of achieving this are as follows:

a choose a strong stock headed by a three-year-old queen and feed steadily from late March, providing a cake of pollen substitute if necessary. Check for swarm cells every week from mid-April, and when they are first found, make up a two-frame nucleus with a frame of food and a frame having at least one queen cell. Then cut out all remaining queen cells, shaking and brushing the bees off each frame in turn to make absolutely sure that none is missed. Normally a second (and larger) crop of queen cells will be found a week later; remove these as before, and also find and destroy the queen. A week later still, remove all queen cells except one good open one, taking care not to shake the comb that this one is on. The stock should now attempt to rear a queen from this cell, but in the meantime a young queen will have emerged in the two-frame nucleus and should have mated by now, so that after another week the queen cell left in the main stock (now sealed) may be removed and the young queen united back to the parent stock, with a minimum of interruption to egg-laying, no swarming and time enough to build up for the main flow with a young queen. Obviously, if anything had happened to the queen being reared in the two-frame nucleus, the other queen cell would be left. The royal jelly may be harvested by cutting down the queen cells almost to the level of contents, flicking out the larvae with a pointed matchstick and then scraping out the jelly and storing it in a small coloured-glass jar kept in the fridge. The handle end of a small teaspoon, if of convenient shape, can serve as the scoop. Each queen cell can contain up to 300 mg of jelly, and by this method up to 30 or 50 queen cells may be produced, yielding in all up to 10 g ($\frac{1}{3}$ oz) of royal jelly. For example, my yield on 30 May 1984 was 2.461 g from 12 cells, an average of 268 mg per cell. The optimum yield is from cells of three-day-old larvae.

b For producing larger amounts a more elaborate technique is needed, as in large-scale queen-rearing. First take an empty brood box and put in two deep frames of honey and two more with large areas of pollen, taken from one or more hives which can spare them; then two frames of open brood plus bees from the colony which is to produce the royal jelly. Arrange the frames so that honey is on the outside, then pollen, with open brood in the centre, and place over a queen-excluder on the brood box of this strong colony. Having first done this, graft young larvae (from any hive) into a frame of about 40 queen cups – either wax or plastic – and place this in the centre, between the brood frames in the top box. Put on a feeder of syrup or super of honey and close up the hive. After three days remove the frame of queen cells, harvest the royal jelly, graft in more day-old larvae and replace. After another three days repeat, this time also exchanging the two frames of brood with two frames of open brood from the box below. The whole process may be repeated for another two or three weeks, by which time something usually happens to prevent continuance. Either the bees swarm out from the bottom box, the queen cups will not be so readily accepted, pollen supplies will not be adequate, or something unforeseen will happen. However, on the basis of 30 cells out of 40 accepted in 8 batches, a yield of 50–70 g (2 oz) of royal jelly may be obtained in three weeks. A Miller-type feeder should be kept on and supplied with syrup throughout, and unless pollen is coming in freely, a patty of artificial pollen pressed into the admission slot of the feeder. Obviously this is also a good way of raising queen cells for queen rearing.

Uses of royal jelly

Although some extravagant claims may at times have been made on its behalf, there is no doubt that royal jelly is an excellent natural food substance extremely rich in protein-building amino acids, vitamins, minerals and many other substances. Doctors in Eastern European countries have found that it is even more effective when mixed with small quantities of propolis and taken in honey. Even when diluted over 10,000 times (45 mg in 450 g), good results have been recorded. A mixture of 96% honey, 3% royal jelly and 1% propolis, taken orally, has been shown to give protection against viral infections. In large-scale tests carried out in Sarajevo during a flu epidemic, 38.8% of patients not given this treatment developed influenza (virus A2), but only 9% of treated patients. Some convincing results on research into claims that it slows down the aging process by stimulating collagen production in the body cells have been reported by Apimondia.

For home use, stir 10 g (about ⅓ oz) into a one-pound jar of warm liquid honey (454 g) and keep as 'honey plus' for the family. This may be used as a winter tonic to be taken in small quantities as required, say a teaspoon (3 or 4 ml) at a time, giving over 100 doses.

Royal jelly keeps well, in fact queen-breeders often keep a supply from one season to another, storing it in a coloured-glass jar, to exclude light, in the fridge.

Eight

Winter – the cosy cluster

THE BEES IN WINTER

Usually a queen two years old or more will have stopped laying by early September, and even a young queen will take a rest from mid-October; feeding in September, or a late minor flow of nectar from the ivy in early October, may bring queens back into lay for a week or more, but in the northern countries of Europe and America brood-rearing is at a minimum from October to December. By the end of December, certainly by mid-January, there will be a modest patch of eggs and larvae in the centre of the cluster and the new bee year will have begun. It used to be thought that queens had a fairly long rest in winter, but on examining colonies accidentally knocked over in December and January I have found areas of brood larger than expected. The work of Dr Jeffree in Scotland has confirmed that this was the rule rather than the exception. (See *The Scottish Beekeeper*, 1956.)

Although bees may withdraw from the end combs by early October, the true winter cluster will not be formed until the first cold spell with night frosts, and even then the cluster is by no means permanent. The winter cluster may be thought of as a hollow shell of bees with the queen in the warm centre, at a temperature never less than about 57°F, but usually over 68°F. If brood is present the warm centre will be 90–98°. Even the outermost layer of bees will keep up a temperature of at least 43°F, and more usually 45–47°F. During colder spells the cluster contracts (to lessen the surface area losing heat), and in mild weather it expands and becomes very loose, with bees going out on cleansing flights around midday and even bringing in pollen as early as the turn of the year on occasion (from laurustinus and helleborus). A site where the midday sun can warm the hives is a great help, as only when the cluster loosens can it flow round a frame to reach other combs of food, and perhaps reform a couple of frames to one side of its first centre. A cluster can move up or down a comb (usually up), or sideways more easily than from one comb to another. This would suggest that in colder areas there might be an advantage in having larger frames, or at least deeper frames, but against this there is the possible advantage of a cluster moving more readily to other frames when food is stored in two boxes, with a bee-space between.

The heat production of an average strong stock at rest in winter is about that of a 4-watt electric light bulb, and the warmth is confined to the cluster, with the air temperature outside the cluster but inside the hive almost the same as the air outside, like a warm bed in a cold bedroom. For all practical purposes the hive income during winter is negligible, and in the natural state a wild colony of bees, established in a hollow tree, for example, has to live on the food stored by it in summer. The bees themselves are programmed to do this, and their winter cluster is designed to do this with maximum efficiency. Anything which breaks the natural routine is harmful to the bees. Any attempt to feed them honey or sugar syrup in winter will excite them and cause them to fly when the temperature is too low, and many bees will be lost. Similarly, any vibration or movement, whether caused by cattle using a hive as a rubbing post, children banging on a hive to see if the bees are still alive, or gale damage, is bad for the bees. At the best it causes untimely activity with increased food consumption, possibly leading to food shortage later on, and it will shorten the lives of the worker bees involved. However, if one is aware of the risk and carries a hive from one part of the garden to another so gently that the occupants are scarcely aware of the movement, a cold spell in January is probably the best time to do so, as after a few weeks when flight is not possible the bees will take more note of new surroundings when they do fly. Any major movement (like loading into a car and driving to a new site) should be postponed until March.

As shown in Table 3, a typical colony needs about 38 lb of food to see it through winter and up to the third week in April, when dandelions usually take over and provide fresh food, both pollen and nectar. In an average year there will be an income in March and early April (from pussy willow, for example), but there is always the possibility of a long spell of bad weather at this time, and a credit balance is better for bees (and less trouble for the beekeeper) than an emergency loan.

No mention has been made of October. This is a very variable month. In some years, hives will increase in weight by up to 8 lb as bees gather pollen and nectar from the ivy, and from late goldenrod and Michaelmas daisy; in other years bad weather will prevent this, but will also cut down food consumption by the bees. This is also the month when 'silent robbing' by wasps, and bees from other hives, can reduce food reserves in a weak stock. For the bees seasonal weather is best, i.e. cold, wintry weather coming in December, January and February (rather than at Easter!).

Bees in mid-winter (15 Nov.–15 Feb.)

In winter one colony uses about 10 lb of honey in 100 days. It may have produced a surplus of 1,000 oz of honey last summer, but by the end of winter its population is down to about 10,000 bees. As a waste product, the colony produces 100,000 drops of water in this time.

Table 3. Typical winter 'balance sheet'

Month	Income (possible)	Expenditure (certain)
Nov./Dec./Jan.	Traces of pollen	10 lb (in 3 months)
February	Some pollen	4 lb
March	0–5 lb (Nectar and pollen)	8 lb
April	0–15 lb (Nectar and pollen)	16 lb
6 months	0–20 lb	38 lb

Table 4. Metabolism of sugars in honey

Each winter day:	$C_6H_{12}O_6$	+	$6\,O_2$	=	$6\,CO_2$	+	$6\,H_2O$	+	energy
	1½–2 oz honey	+	1 cu. ft oxygen	→	1 cu. ft carbon dioxide	+	1,000 drops water (as vapour)	+	3½–4 watts

The equation for the metabolism of sugars in honey is shown in Table 4.

Honey contains about 18% water. Air contains 20% oxygen. Bees cannot extract all the oxygen from any given amount of air, so in practice need four or five times this amount. To provide this, the air in a hive in winter has to be changed about once every twelve hours by slow, steady diffusion and ventilation.

Top ventilation

In the old days many people thought it best to 'keep them warm in winter' by piling on old blankets, sacks, sheets of newspaper, old trousers, etc., and many stocks were lost as this packing became saturated from condensation, even dripping water back on to the bees. Nowadays most beekeepers appreciate that the weekly ¾ lb honey consumed in winter produces about a cupful of water (in the form of vapour), and this has to get out of the hive to keep the bees reasonably dry. So it is better to have no packing at all on top, just an open feed-hole in the crown board, with an air space under a secure ventilated roof. An empty super (no combs) over the crown board has many advantages, also giving space for a circular feeder or block of candy. So long as the bees are clustered, there is no question of keeping the whole hive warm (on a power output of approximately 4 watts), only the actual cluster itself. At Buckfast Abbey they put thin wedges of wood or nails between crown board and brood box at the back to give top ventilation. One or two elderly beekeepers I know (with equally elderly hives) still use blankets on top and get good results, not realizing that they are unwittingly providing ventilation anyway by the cracks and crevices in ill-fitting hive parts.

To complete the picture, at the risk of confusing some, it should be said that from mid-February to mid-April the need for water usually exceeds the production of waste water, and bees then have to bring water into the hive to dilute the stored honey to feed the larvae. Any condensed water is then lapped up and recycled, as in modern submarines. Heat insulation provided by a blanket of glass wool or a square of white polystyrene on top of the crown board may be of some use then as the brood nest is growing more rapidly than the number of adult bees, who are hard pressed to cover the brood.

BBKA winter survey

In September 1979 the British Beekeepers' Association published a winter survey, based on the experience of the previous six years. As this was compiled from reports voluntarily submitted by co-operative beekeepers, one may suppose that the other kind had rather more losses and preferred not to say anything, or perhaps just forgot. People who can forget to feed their bees are quite capable of forgetting to fill in and post a form. Anyway, this survey disclosed that 4% of stocks died during winter and another 4% survived but were queenless. The biggest single cause of death was starvation, with diseases like nosema, acarine and amoebic dysentery (in that order) coming next. Mouse damage, wax moth, woodpeckers all contributed, and not mentioned at all were the two isolated cases of badger damage in one recent year. According to the survey, 80% of all beekeepers now accept that top ventilation in winter is best (as opposed to the 'keep them warm with old blankets' school of thought).

The probable truth, in my opinion, is that winter losses may vary from 2% in a good year to something like 15% or even more in a bad year, with mouse damage coming a close second to starvation, followed by queen failure and disease. In a really hard winter (like 1963) damage by woodpeckers can be a real trouble.

Hazards of winter

Most of these can be guarded against, but accidents can happen:

Gales

Gales can blow off a hive roof, but a couple of bricks or a concrete block on the top will prevent this. So will a piece of rope or wire passed around the hive and tied. A bough falling from a tree can knock the hive askew, and here prevention is better than cure – make sure the hives are where this cannot happen.

Bullocks

These and other farm animals can get into the apiary and scratch their hides against the hives, as if they were trees. This can knock them over or displace a box from a floor and let the rain drive in. So fence or wire the apiary securely.

Vandalism

Vandals can push a hive over, or pelt it with rocks, bringing out the bees when it is too cold for normal flight. However, warm bees from the centre of the winter cluster can fly and sting within half a minute of the colony being disturbed, although the activity caused and food consumed impose a penalty: the life of the bees is reduced and more food is consumed in the next two hours than in a quiet week.

Mice

Mice, which would not dare to enter in summer, can get in when the bees have clustered tightly in cold weather. Once in, they will eat into sealed food combs and build a nest of grass and leaves in which to breed, thus protecting themselves from the bees when a warmer spell allows them to break cluster and move around. In winter a sign that all is well is a sprinkling of fine wax dust at the hive entrance, as the bees chew off the cell cappings to get at their winter food. Larger pieces, the size of fingernail clippings or bigger, indicate the presence of mice chewing down the food combs. Simple but effective mouseguards can be made from scrap expanded metal (diamond mesh) found on building sites, cut into strips and tacked across hive entrances in October. Best of all is a permanent, built-in mouseguard (see page 156).

Queenlessness

An old queen may die, or fail to lay fertile eggs, before the coming of spring. The bees can do nothing about this, unless the queen is alive and fertile until April at least, when they just might raise a new queen. Sometimes the colony will die out, but it will often survive until the first inspection in March, when a comb of eggs and brood from a hive close by can be given as first aid. Even if no queen is successfully reared, the hive population is then balanced and reinforced. The real answer is to re-queen any doubtful stock by mid-September.

Disease

Any latent disease, such as nosema, for example, can weaken a colony to the extent that they either die or are too weak to be worth saving in April. Remedies are various and are best applied in summer and autumn. Nosema can also be treated in spring (see page 26).

Winter – THE COSY CLUSTER

Starvation

This is the major hazard. A stock which was fed in September may be short of food before the end of winter because of 'silent robbing' in the autumn, by wasps or other bees. The remedy is candy, described in detail on page 167.

Woodpeckers

Woodpeckers can peck large holes in the sides of hives in order to get at the food combs. This is not just a new trick (like blue-tits and milk bottle tops); after all, wild bees live in trees and in the natural state woodpeckers must often have found their combs and learned to appreciate their value. A 6-ft length of wire netting wrapped around the hive will effectively prevent this trouble.

Wax moths

These are not usually a problem when bees are strong, and in any case they are killed off by frost, so the best advice is to store supers or spare combs under a sound roof out of doors.

BEEKEEPING THOUGHTS IN WINTER

In spring, summer and autumn beekeepers are too busy with bees, gardens, sport and the outdoor life generally, to sit down and think out the basic logic behind their beekeeping practices.

In winter there is time to think and discuss with others, and the purpose of this section is to stimulate constructive thinking.

Management

Now that honey has been extracted, the bees fed where necessary and hives prepared for winter, there are a few months in which one can consider how to do better next year. The most important factor of all, the weather, lies completely outside our control.

The second most important factor, having good forage for the bees, is only partly under control, in that most of us have to keep bees either where we live, or within a few miles of home. Fortunately it is possible to produce a surplus in most areas of Britain; in fact, urban and suburban areas are much better than they were fifty years ago. Against this, much of the farmland in the Midlands and the eastern half of England is less productive of honey, with larger fields, fewer hedgerows, and weed-free prairies of barley and wheat. However, the rapid growth in oil-seed rape acreage (bright yellow blossom), and to a lesser extent newer types of field beans, are redressing the imbalance. Some factors lie entirely within our control, namely the queen and the type or size of hive.

Queens

Probably the most important single factor in honey production is the queen, who may be likened to the mainspring of a clock, or perhaps nowadays one should say the battery of a watch or clock. The normal productive life-cycle of a queen goes something like this. *a* Her first year, in which she rapidly builds up to peak performance, but unless reared and introduced very early in the year (or in September of the previous year), cannot mother a colony strong enough to exploit the summer honey flow to the full. In this first year the likelihood of swarming is usually low, possibly only 5%. *b* A second year in which she probably gives her best performance, but is rather more likely to swarm (perhaps 15–20% chance). *c* A third year in which performances usually (but not always) decreases a great deal and swarming is much more likely (35%-plus). *d* A fourth year of quiet, restful life. Egg-laying much reduced, hive population small, bees seem content but produce no surplus. Very likely to swarm and may fail altogether next winter and leave a queenless hive at a time when nothing can be done about it. The colony either dies out in winter or lingers on, very weak, until spring. Weak stocks also have less resistance to disease. Another possibility is that the old queen may become a drone-layer, or at least produce far more drones than are necessary for the well-being of the colony.

This is, of course, a generalized picture. Sometimes the bees will supersede, i.e. replace the failing queen, by building just one or two queen cells (usually nearer the centre of a comb than swarm cells are). Some beekeepers may boast of their 'leave them alone' methods and claim large honey yields. They may even get them, sometimes, but the norm would be very much less than that gained by close management.

It may help to look at what the big professional beekeepers do, especially overseas where honey is produced in really large quantities. I know of one commercial beekeeper in New Zealand who re-queens every hive every year, buying in young queens from a queen-rearing specialist, because he works single-handed and says he hasn't time to rear his own. A very large honey business in North Island, New Zealand (7,000-plus hives) re-queens every hive once in two years. I helped them do this recently at the end of February and early March, after the honey crop had been taken off. This corresponds to end of August and early September in Britain, and means that the stocks have a young queen to see them through the winter, and that she is still in her first year throughout the next summer. She has a second season and is then replaced. This honey-producing firm regards nine-day inspections and any swarm control methods as far too expensive in labour, and accepts that a small percentage will swarm. Their apiaries are mostly well away from any built-up areas, often with no trees large enough to provide natural homes for swarms, and so it is possible that most of the swarming workers will drift back to the apiary anyway. So, my first firm recommendation is to think about re-queening once in two years, as described in an earlier section.

Brood boxes

Most beekeepers in Britain (over 85%) keep bees on British National frames each having a comb area of about 90 square inches, i.e. 180 square inches counting both sides of the comb. The question to decide is what area of comb should be allowed for the queen to breed in, so that a good queen has enough room yet not too much, or else the honey harvest may be stored below the queen-excluder where the beekeeper cannot easily get at it, and would not usually wish to do so, as honey from the nursery area is not thought appropropriate for human consumption. The choice is between three possibilities – single, one and a half or double brood-box management.

Egg-laying

The basis of all calculations on comb area needed by queens is the fact that 21 days elapse from the laying of the egg to the emergence of a worker, and usually another couple of days may go by before a cell has been cleaned up and polished by young bees, and the rim of the cell chewed smooth where emergent youngsters have left a slightly ragged edge. If a queen can normally lay in a cell again after 23 days, then we have to think in terms of the maximum number of worker eggs a queen will lay in 23 days, and provide enough cell space for this, plus provision for the inevitable drone brood, and for pollen and honey stored close to the actual brood nest. Many beekeepers (and writers of bee books) speak glibly of modern, prolific queens laying over 2,000 eggs a day. The important statistic is not the *highest daily figure* achieved but rather the *highest daily average sustained for 23 days*.

A thorough piece of recent research involving over 50 stocks showed that the best queen out of 50 laid an average of only 1562 eggs per day over a 23-day period, although many queens did reach a peak of over 2,000 for several days. The average 'good queen' laid around 1200 to 1400 per day over the trial period.

Considering the best queen (1562 eggs per day), her cell requirements are $23 \times 1562 = 35{,}926$, say 36,000 for brood – a comb area of 1440 square inches, provided by 8 National frames. However, it is usually considered that another 5 or 6 National frames are needed to accommodate pollen and honey reserves close to the brood area, and to allow for imperfections in the comb itself, as well as for areas of drone brood. So, it is argued, the equivalent of about 13 or 14 National frames is needed in brood box or boxes, and current fashion decrees that a single National (or Smith, or WBC) brood chamber is not enough. Thus it comes about that some beekeepers use 2 brood boxes per hive (22 deep frames), most use $1\frac{1}{2}$ boxes (a deep and a shallow) equivalent to about 16 deep frames, and a few use a different, larger hive.

Disadvantages of 2 and $1\frac{1}{2}$ brood box systems

First and foremost, the main practical disadvantage is that there are 22 frames involved, and in the case of $1\frac{1}{2}$ boxes they are even of different sizes and not interchangeable. Secondly, most of the extra space provided is not needed, or perhaps only needed for three or four weeks in the year. In the case of a double brood box hive (larger than a Dadant, even), either the outer combs are not used from one end of the season to another, or else the brood is found in the upper box after the first two years, and the lower box becomes a reservoir of old, black and gnarled combs. Thirdly, the honey gathered by the bees is first stored in the deep boxes, and in a poor season there is no surplus available for the beekeeper. What is the alternative? It is to use a single brood box plus a super always kept immediately above the queen-excluder as a food reservoir for winter, but capable of having honey extracted from it, as it has not been bred in.

Single brood chamber management

For several years the writer has kept one apiary each of the larger hives, British Commercial and Dadant, and three apiaries of British National hives on single brood chambers. In a good year (like 1983) they have all done very well, but in an average year the National hives have yielded more, simply because the honey has been put in the super or supers where the beekeeper can take it off and extract it. In practice it is noted that the bees regard the central combs of the first super as part of their brood nest, disregarding the queen-excluder in between, and store a great deal of pollen in them. Pollen remains in the combs after honey has been spun out, and so the honey content of the first super is rather less than expected. However, the pollen can be scraped out with an old kitchen spoon, thoroughly mixed with fresh honey and marketed as 'natural pollen preserved in honey'. This leaves the wax midrib on which cells can readily be drawn out again next spring. Under favourable conditions bees gather a surplus of both pollen and honey, and it is as logical to harvest one as the other. In any case, pollen not used by the bees within a few months deteriorates, which of course honey does not. In fact, any pollen not capped or within the brood nest area will go mouldy and have to be ejected by the bees next

April when the brood nest expands very fast.

It is sometimes said that queen-excluders should be removed in winter as they are a barrier, and the cluster might get separated from their queen. In practice this very seldom happens. In a double brood-box system the food reserves tend to be more on the outer combs, and in such a large space it has been known for the cluster to move to one side and to have used up those combs of food just as a cold spell strikes in February. The bees then die of starvation a foot away from food on the other side, because it is too cold for them to move across. When most of the food is above the cluster, movement is easier.

Conclusion

Few will dispute the wisdom of replacing two-year-old queens. Advocacy of a single National brood box will cause many eyebrows to be raised, but nevertheless some of the most successful beekeepers in Britain use this system, professionals as well as amateurs. Two factors are of paramount importance:

a maintenance of brood combs to a high standard, so that all cells can be used by the queen;

b keeping one super over the brood chamber as a reservoir of food for the winter, since there is not normally room for 38 lb of winter food and the bees in a single box.

STOCKPILING MATERIALS

With practice, one can develop a 'scrounger's eye' for useful raw materials, obtainable very cheaply or at no cost. The point is that they have to be stockpiled when available, ahead of the actual need for them. In many towns there will be a builders' yard where material from demolished houses is sold, and this can be a source of well-seasoned timber from old floor-boards and joists. There will be nails present, often hidden, and care is necessary if using a motorized saw, but some fine supers and floor boards can be made from such wood. Save plastic 1- and 2-gallon screw-top containers, as these will be useful for holding and transporting sugar syrup for feeding in the autumn. Look out for old hessian sacking – the more rotten it is the better it will burn in your smoker, after drying in a greenhouse or shed. One of the few remaining sources of this material in this plastic age is your local health-food store – make contact and ask them to save their hessian bags for you, leaving a pound of honey to establish your identity as a beekeeper. This may also lead to a useful retail outlet later on. Get on good terms with your local baker; they have very strong plastic buckets (with lids) which make excellent bulk containers for honey. Once friendly relationships have been established they may be able to supply baker's fondant at trade prices, and this is a perfectly adequate substitute for bee candy. Your friendly butcher will have empty plastic containers which have held frozen New Zealand lamb's liver. These make excellent feeders with very little work, as shown in another section. As your personal contribution to the 'Keep Britain Tidy' campaign, remove unsightly concrete blocks left behind on grass verges, perhaps a surplus from work some months earlier or fallen off a lorry! Rusty angle-irons can be rescued from a dump, wire-brushed and painted with black bituminous paint; a local school near my home recently demolished some old cycle sheds, and the vertical irons into which the front wheels had fitted made wonderful hive stands, about 6 ft (2 m) long, in U cross-section, for resting on concrete blocks. A building site can furnish offcuts of diamond mesh expanded metal for use as mouseguards. From old, worn carpets cut squares or rectangles to the size of your hives; these will be very useful as temporary covers or bee-proof quilts in the apiary. Large waste pieces of good hardboard, used in advertisement displays by stores and put out at 9 a.m. on 'dustbin days', can be had for the asking. Cut into 18-inch squares, or to your hive size, these take up very little storage room and will make crown boards later on, for nucs as well as hives. If given pigeons or pheasants for the table, save and dry the wings complete for use as bee brushes; a large goose wing-feather will be similarly useful. Sometimes fellow beekeepers have spare, old WBC equipment for sale, or even to be had for the removal. As explained on page 157, an excellent solar wax extractor may be made from a WBC lift, and a double nucleus from a WBC brood box.

WORK FOR BEEKEEPERS IN WINTER

Beginners

Those who perhaps started up with a swarm or nucleus last June, should in winter be planning their garden or home apiary to accommodate two hives on a stand, with room between for a nucleus later on. Those fortunate enough to have a large garden can easily find a corner for a couple of hives, but most are not so lucky, and it is necessary to plan very carefully, so that the bees may not be a nuisance to neighbours. As mentioned earlier, the important factor here is that there should be some form of screening about 2 m ($6\frac{1}{2}$ ft) high to make the bees fly upwards, and not cruise at chest level across a path or neighbour's garden. This may be done by using a quick-growing, close-trimmed hedge, a woven or larch-lap fence, or even a tall crop such as runner beans or artichokes. Although these crops will not form an effective barrier until late June, neighbour trouble, if it comes at all, tends to come in July or August when bee populations are highest and sunbeds most in use!

Most beginners do not wish to have more than two hives, to provide honey for their families and friends and perhaps Christmas presents. With the addition of one or two nucleus boxes, on the same stand, this can provide plenty of scope for more advanced work on bees without consuming very much time. The pattern shown in Fig. 76 has many advantages and is easy to make, even without a workshop. The important consideration is to progress to two hives as soon as possible. With a single hive one is vulnerable, and should a queen fail at any time there is no simple solution, whereas a frame of eggs from another hive in summer would have solved the problem easily. One task for the first winter is to make a second floor, so that the existing colony can be given a clean, dry floor in March as part of 'spring cleaning'. Winter evenings can afford time to attend lectures arranged by the local beekeepers, or technical college, and of course to read one or two books: a suggested reading list is given later on. If you are now in your second year as a beekeeper, you should be planning to take the BBKA preliminary exam next summer. It is entirely practical and oral. There may be winter classes already arranged to cover what you need to know; if not, most local secretaries would be only too pleased to arrange for a local expert to give a talk to a group keen to take this exam next summer.

Apart from buying a second nucleus or taking a swarm, the simplest method of increase is by making an artificial swarm in June. You may need help with this, but at least you should read up about the technique and ask a lecturer or experienced beekeeper to explain it at a winter meeting, and plan to acquire a second hive in good time. A reasonable target would be to make the floor, crown board and roof, and buy the actual brood box.

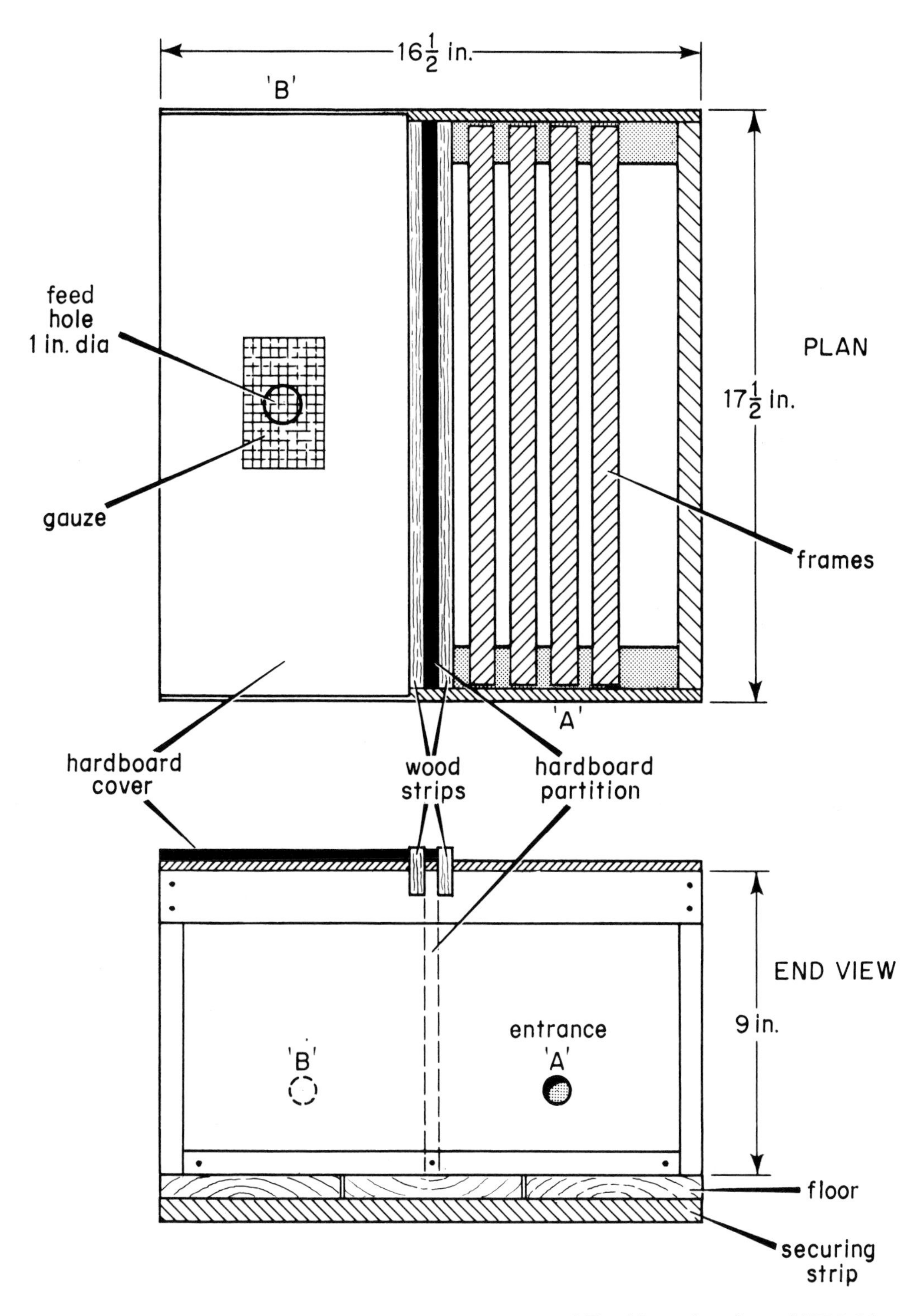

76 Double nucleus from old WBC box.

Winter – THE COSY CLUSTER

Improvers

An 'improver' is defined as a beekeeper with two years' experience, and in this book it is assumed that one spends three years as an 'improver', after which time the term 'experienced' is used to cover the years that remain, whether the beekeeper wishes to expand to commercial work and try new techniques or just carry on happily with a couple of hives. Even for the latter, there is no doubt that the stimulus of taking an exam in one's hobby sharpens the interest considerably, and ensures that a reasonable level of competence is reached as soon as possible. So, if that BBKA preliminary exam was not taken last summer, plan to take it next year.

If you have decided to go on as a genuine 'improver', then winter is the time to plan ahead and make equipment which you will need. The first should be a nucleus box, and Fig. 76 shows a diagram of a double nucleus box made from an old WBC brood box, often obtainable very cheaply second-hand. Details of making nuclei are given in Chapter 5, and this should be regarded as the most important single task this year. If you have no workshop at home, there are evening classes in woodwork at local schools or technical colleges, and you will find that the instructor will give every help and encouragement. Hives are extremely expensive to buy, and a few pleasant evenings at the local 'tech' should provide a couple of hives as well as a nucleus box.

In winter your reading programme should be extended, and it is suggested that in three years as an improver some of the following techniques be chosen, and the exact procedure carefully studied and noted at leisure.

Suggested techniques and activities for improvers

1. Work towards BBKA preliminary exam (if not already taken).
2. Queen introduction.
3. Honey-handling (storage, warming, bottling).
4. Hygiene (blowlamping boxes, sterilizing combs).
5. Preparing hives for the heather.
6. Setting up a small out-apiary.
7. Pagden method of hiving a swarm.
8. Setting up a two-queen colony.
9. The Taranov artificial swarm method.
10. Comb honey production.
11. Honey and wax as show exhibits.
12. A simple method of queen-raising on a small scale.
13. Making wax foundation and other beeswax products, e.g. cosmetics, furniture polish.
14. Making mead.

Experienced beekeepers

There are no limits to the field open to an 'experienced' beekeeper, apart from those of time and space, energy and choice. The development of some systematic queen-rearing practice could well go hand-in-hand with the production of nucs and queens for sale. The construction of equipment for this, and for the extension to any desired or practicable number of hives, in units of 10–20 per apiary, with carefully planned premises for honey-handling, storage of equipment, wax-handling and foundation-making, would certainly justify a shed or end of garage converted to a handy workshop. Any major provision like this is best made in winter, when bees and garden make less claim on one's time.

By careful management it is possible to obtain up to 1 lb of beeswax per hive per year, and with anything over 6–10 hives this opens up a new field, not only to make all the wax foundation needed for good comb maintenance, but also for production and sale of high-quality furniture polish, cosmetics and 1-oz (28-g) bars of beeswax.

If an experienced beekeeper wishes to sell his products to best advantage, then running a bee stall at fêtes and shows is definitely the best way of doing so, and a great drawer of crowds if you display a glass observation hive, holding two brood combs one above the other with queen, drones and workers all clearly visible.

MAKING YOUR OWN EQUIPMENT

A double nucleus box

A sound WBC brood box can be made into a very useful double nucleus; so can a National brood box, of course, but in my experience there is so much more old WBC equipment to be found, and often had for the asking, that it is more economical to use it. It is also much lighter.

First make a hive floor approximately 17 in by 16½ in, but to fit your actual brood box. This can be done from three strips of salvaged timber, preferably at least ¾ in thick, held together by a couple of flat strips nailed across two opposite ends, underneath. Since all WBC equipment has a bottom bee-space, with the tops of the frames lying flush with the top of the box, it is necessary to tack ¼-inch-thick strips of wood all round the top of the brood box, so that a couple of crown boards can rest on the box with a bee-space under them, to avoid crushing bees.

Now rest the box on its floor and measure up for the hardboard or plywood partition, normally 14½ in by 9½ in, but there may be small individual differences, so work from your actual box. The partition has to fit flush on the floor and stand about ¼ in proud of the top,

77 Special entrance block, serving as a mouseguard in winter.

being held in place between two wooden strips 17 in by $\frac{7}{8}$ in by $\frac{1}{2}$ in nailed to the top of the brood box, and by tacks or thin wooden strips nailed to the floor. Use putty or some filling compound to block any chinks in the division wall between the two nucs. Drill two entrance holes $\frac{3}{4}$ inch diameter in opposite ends of the box and a 1-inch diameter feed hole in each crown board, cut from hardboard or plywood to measure approximately $17\frac{1}{2}$ in by $7\frac{1}{2}$ in, but made to fit your own space. Now fasten the floor to the body of the hive, preferably with screws. As there is no question of changing floors, the fitting is permanent. Each nucleus will take four frames comfortably, with room for a spacer dummy board if desired. An old WBC super will fit on top to accommodate two small, round feeders. If used only during summer months a temporary, improvised roof will do, but it is better to use a normal hive roof. A pair of four-frame nuclei with young queens will come through winter very well, so long as they have enough food and a dry roof. These last two winters I have been away in New Zealand from late January to the third week in March, and put 2-lb slabs of candy on all nucs before leaving. In no case was more than half the candy eaten, and all nucs came through well, six one year and seven the next.

An improved entrance block

In Central Africa some thirty years ago I was plagued with large black-and-orange flying beetles invading my hives during the dry season. Adansonii bees are not very docile at their best, and these apparently sting-proof invaders made things worse. To keep them out I devised a palisade of rounds nails driven into the entrance block at 9-mm centres; now back in Britain for the last twenty years I find the same device a very effective mouseguard. With the addition of a long screw driven up through the floor as a hinge pin, the entrance with a built-in mouseguard is always there when wanted, and is absolutely no impediment to pollen loads, drones or queens. It can be pulled open during the main honey flow, and pushed in at the end of July to restrict the size of entrance to be defended against wasps (and other bees). I used to winter on full-width entrances, but more recently have used top ventilation with these blocks in position all winter, in fact for most of the year except for a few weeks in high summer. The point of having the nails sited two-thirds of the distance in from the front is to give room for a wedge of foam rubber to be pushed in temporarily if moving the hive. The choice of 9-mm centres (giving about 8 mm actual gap) is based on the size of a mouse's skull, to exclude these pests in winter, and the fact that the nails are vertical means that no pollen loads are accidentally knocked off in spring, which can happen with the round holes of the orthodox mouseguards. Above all, the convenience of having a permanent mouseguard means that the task of visiting an apiary in October to fix guards is eliminated.

The best way to get a regular spacing of nails is to use a template made from a thin metal strip drilled at 9-mm centres, and to drill the block with the template clamped to it, the same clamp holding the block firm to bench or table. It is best to cut off the heads of the nails and then tap them right into the wood. If adapting an existing floor, it is best to shorten one side-piece so that the entrance block fits as shown in Fig. 78. This obviates the nuisance of a block which can get pushed too far in, or is difficult to prise open. If making a floor, then the one-inch-square strips on which the brood box rests should be cut in four lengths, each one inch shorter than the standard side of the hive, and the entrance block made from one of them.

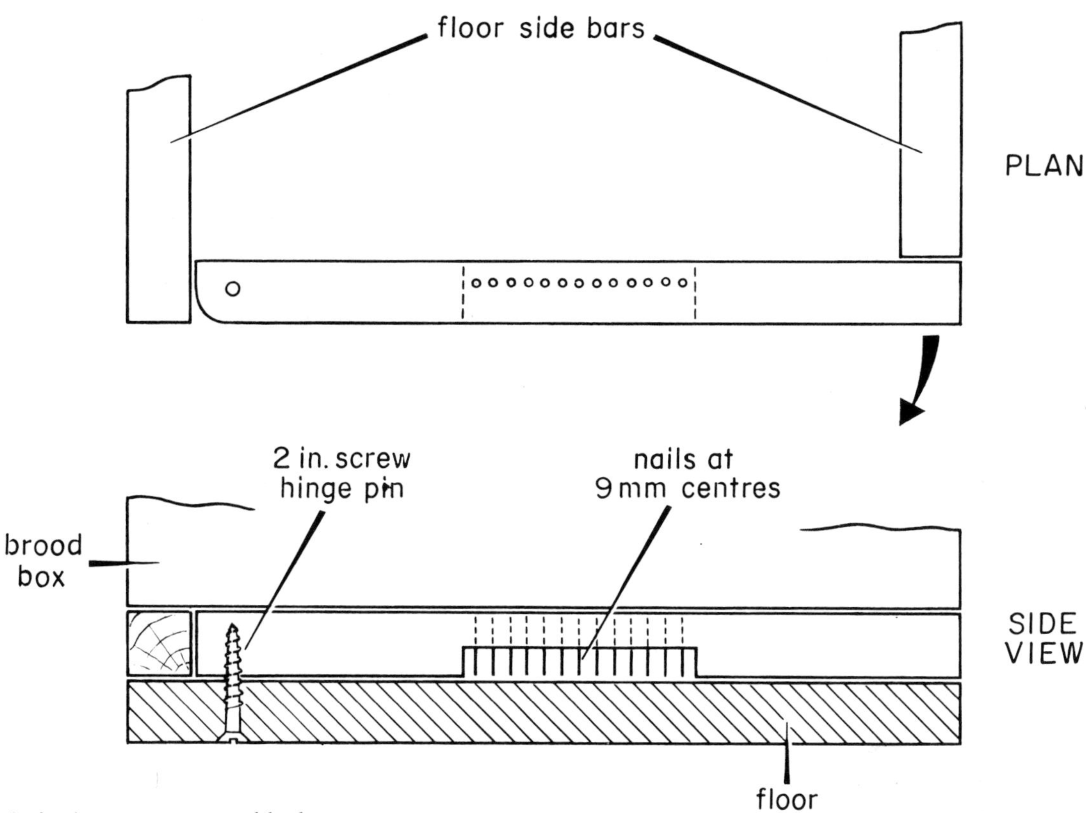

78 Anti-mouse entrance block.

A solar wax extractor

All conventional designs for solar extractors are double-glazed, expensive and easily broken. Yet one sheet of *thick* plastic is as effective as two sheets of glass and much more durable. Fig. 79 illustrates a simple model made from an old WBC lift, with a base of thick timber or exterior ply, having a layer of insulation (glass wool or polystyrene). Inside the box is a shaped, sloping surface of fairly heavy metal (made from an old iron oven top, hive roof or similar scrap), to carry the wax fragments and old combs. A bent strip of coarse gauze will hold back unmelted wax and most of the dross or 'slum-gum'. One of the most useful and surprising features is the way that clear wax runs away from the rubbish, which tends to stick to the metal sheet, and should be scraped off in the evening while still warm. Old, black brood combs should be broken up, soaked in rainwater for 24 hours and boiled in water in an old hessian sack weighted down with stones, in a clean drum over a garden fire, as shown in Fig. 80.

QUEEN-EXCLUDERS

Their purpose

As everyone interested in beekeeping will know, queen-excluders are designed to allow worker bees free access to the honey supers while confining the queen (and drones) to the brood box or boxes: this separates the nursery

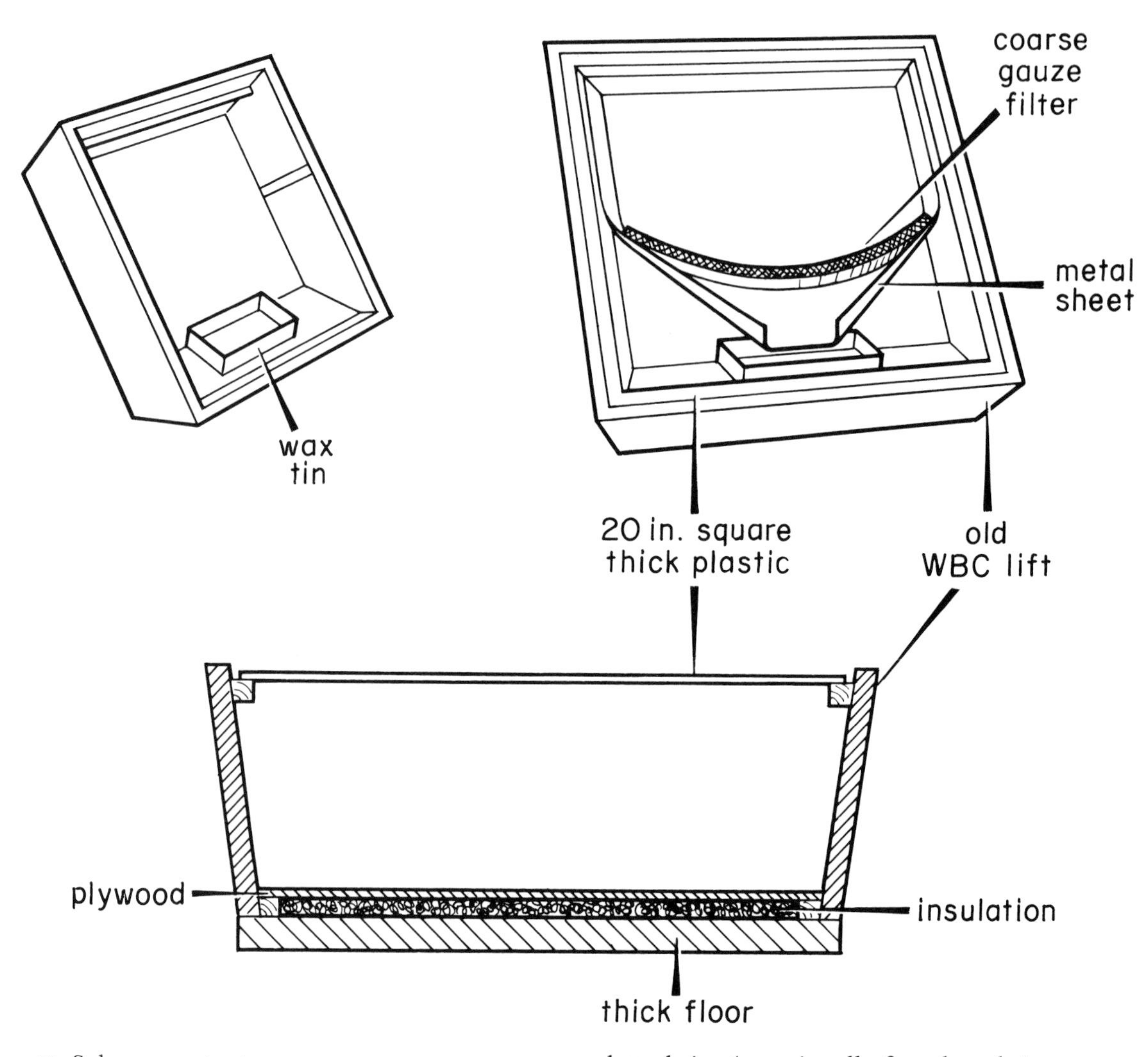

79 Solar wax extractor.

and colony storeroom from the owner's pantry above. In wild colonies the stored honey is usually found above or on the outer combs, but can be found below mixed up with brood combs. In the days of skeps the beekeepers often had grubs, pollen and dead bees mixed with their honey: nutritious though the mixture may have been, it would not appeal to many people nowadays. There are two main types of excluder in use in Britain today: the slotted zinc (occasionally found made in strong plastic) and the Waldron wire. A German wire excluder (the Hertzog) is much stronger than the Waldron, and is gaining in popularity, either framed or unframed. All have gaps of 0.165 in (4.2 mm). We owe the original invention to the Abbé Colin (France, 1849).

The slotted zinc excluder

Four out of five amateurs use this type, which rests directly on the tops of the brood frames, so that not much more than a quarter of the area

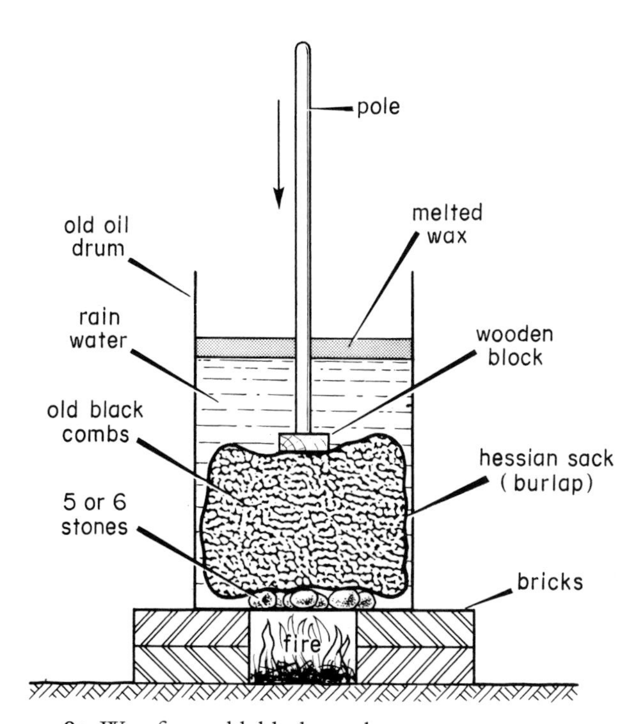

80 Wax from old, black combs.

of the excluder is available to the bees. Should the excluder be placed with the slots parallel to the frames, then the available area is still further reduced, as every other frame top blocks a fraction of one complete line of slots. Fortunately 10% of the area of an excluder is ample for the worker's needs, but of course the beekeeper is paying for 100% of that area. Much more annoying even than this is the fact that the excluder gets stuck down with propolis, and even if carefully peeled back in the orthodox manner, disturbs the bees unnecessarily, and sometimes it is not too easy to straighten out the zinc before replacing it. Also, in scraping the surface to clean it, a small piece of metal can get caught up and very slightly distorted, leaving a space through which a slim queen can squeeze. Another practical difficulty is that the bees tend to force propolis between the zinc and the frame tops, pushing up the flexible excluder so that the base of a frame cannot get down to its full depth, and the frame shoulders are not resting properly on the rebated tops of the super. If this type of excluder is inadvertently used on a hive with a top bee-space, then it sags in the centre and the space between the zinc and the frame tops varies from 8 mm ($\frac{5}{16}$ inch) at the sides to nothing at all in the middle. The excluder is then stuck down over a considerable area with propolis and burr comb. We are often advised to frame all excluders, but with an $18\frac{1}{8}$-inch-square slotted zinc excluder this is only effective if reinforcing slats are fixed at 6-inch centres in addition to the outer square framing. Short-slot zinc excluders are slightly stronger than long-slot ones, but both should have been phased out years ago.

The Waldron

In the writer's experience about 5–8% of beekeepers still use the more expensive British framed excluder with rigid wires. For some inexplicable reason they are not framed with bee-space on one side only as they should be, but with half a bee-space on either side. The typically British compromise unfortunately gives us the worst of both worlds, yet is manufactured and sold to this day, as it has been for nearly a century. This type of excluder, as one would expect, normally collects a good deal of burr comb, also cell comb extending down from frames, but at least is so strongly made that it can be levered off without distortion. However, more and more beekeepers are now using the Hertzog, also a wired excluder but much more strongly made. Some year ago this could only be obtained direct from Germany, but it is now stocked by dealers in Britain.

What is needed?

Wild colonies have often been found in fairly small spaces, with access to another space via a hole of a square inch or less, and this barrier has not hindered the bees from building ample combs beyond it. In the days of skeps, supers (or 'caps') were frequently placed over a round hole only 2 in (50 mm) across. In Central Africa, with no access to beekeeping appliance stores, I managed perfectly well with one old zinc excluder cut into 16 pieces, each roughly tacked over a hole cut in a square board. Supers placed over such excluder boards regularly gave yields of over 100 lb (45 kg) a stock per year.

Clearly, what is needed is a rigid board framed to give a bee-space of about $\frac{5}{16}$ inch on one side, leaving the other side flush over the whole area. An insert of about one-fifth or one-sixth of the area of the board has been found entirely adequate, and zinc of this width (about $3\frac{1}{2}$ inches) will not sag or bend easily, while the provision of a bee-space over the frames will give the workers access to every slot or gap in the actual excluder.

81 Queen-excluder with bee-space on one side only.

Making five from one

An old slotted zinc excluder, discarded perhaps because of two or three faulty slots, may usually be cut up into at least three or four suitable strips. A sound one, even of the smaller WBC type, usually gives five, and each strip should be fastened to cover a slightly smaller gap cut into a framed plywood board, as shown in Fig. 81. At first I just tacked them on quite roughly, but one exhibited at the 1976 Devon County Show was held down by a number of very small screws neatly arranged. Current practice is to stick the zinc down with a liberal application of Cascamite. The bees seem to approve, and usually add propolis along the edge of contact between zinc and wood, to make the join even more effective.

A somewhat different technique is used with strips of old wire excluder cut from Waldrons; these strips are tacked across the centre of square framing $\frac{5}{16}$ in thick and then plywood strips on either side fastened to complete the job. By tapping the plywood sideways on to the sharp end left when the excluder is cut up, a firm joint is made laterally, which the bees improve by propolizing when on the hive. In practice the writer usually puts these excluders on so that the slots or wires are at right angles to the brood frames below, so that bees from all frames have equal access to the supers. Experience over many years, confirmed in 1983 when hives filled up to five supers over such excluders, has shown that a limited area of excluder certainly does not affect the ability of bees to pass through freely. In addition I consider brood nests are somewhat 'cosier' in March and April with a solid covering over four-fifths of the area, yet with very adequate through-ventilation from November to the end of February, when this consideration is more important than 'cosiness'. Beekeepers who hesitate to winter colonies with just a full-size excluder (and of course a roof) over the cluster, may feel that they may more safely do so with the modified type described. Why not save money and provide a better excluder by making up some this winter?

The ease with which full supers may be separated from these excluders when taking off the harvest was well demonstrated in 1983 when I still had a few slotted zinc excluders in use. The contrast was so marked that groans and imprecations were loud as the zinc excluders stuck and lifted, while the supers above board excluders separated easily, with no disturbance to the bees.

In countries like New Zealand, with larger and more consistent honey flows, commercial beekeepers usually do not use queen-excluders, and the pressure of honey stored above pushes the queen down to the brood area. Unfortunately our honey flows in Britain are less dependable, and four years out of five a queen given free access to all boxes would 'chimney' up the middle and brood would be found mixed with the honey in the supers.

MAKING PRODUCTS FOR SALE

Furniture polish

Two parts by weight of white spirit to one of beeswax melted together and stirred will make ordinary polish, but a superior version, suitable for antiques, can be made by stirring 10 oz of beeswax shavings into 1 pint of pure natural turpentine, heated to about 160°F/71°C in a double saucepan. Under no circumstances may turpentine be heated over an open flame. When the beeswax has completely dissolved, pour the mixture into 120 ml flat tins.

Cosmetic creams

A basic recipe would be 15% beeswax (of a light colour), 55% of medicinal paraffin (possibly including a small proportion of almond oil), just under 30% rainwater and slightly less than 1% of borax, plus a drop or two of perfume. The wax should be grated into the warm paraffin in a double saucepan. The borax is separately dissolved in warm rainwater and stirred into the other mixture while both are at a temperature of about 160°F. Stir as it cools, adding the perfume at 140°F, and pour into jars (previously warmed) at 120°F.

ATTENDING BEEKEEPERS' MEETINGS

It really is important to get together with other beekeepers, of whom there are over 40,000 in Britain: nearly one in every 1,000 of the population. Wherever you live, there will be a branch of a county beekeepers' association within a few miles. Here you will find the obvious advantages of local know-how, getting help from others more experienced, obtaining bees in the first place, and sharing facilities such as expensive extractor and honey tank, club library books, and winter lecture programme. There is also the bonus of friendly social contact with people from all walks of life, sharing a common interest.

A whole new field of interest is provided by microscope work, not only for disease diagnosis, but also for pollen identification and closer study of the anatomy of honey bees. This demands many hours of practice and thought, but is well worth while.

Study groups

After the first season it would be wise to think of taking the preliminary examination of the BBKA, and many associations plan a winter lecture programme aimed specifically at this. Further examinations such as the intermediate and senior exams of the BBKA provide the motivation for serious study which will feed back into better beekeeping and more honey, as well as leading to a mastery of the craft and great personal satisfaction. For the intermediate exam, the best possible preparation is to form a study group of 6–8 like-minded colleagues and arrange to meet one evening a week at each other's houses. We did this most successfully in Torquay and I enjoyed fostering a similar enterprise in Bristol, which was equally popular.

Nine

The New Year – health checks and bee care

BEGINNERS – JANUARY

During January and February a well-stocked hive should need no attention, but an occasional check after a gale might be wise, especially if your hive is away from home. A roof may blow off or a falling branch knock the hive askew; cattle may push through a hedge and use your hive as a scratching post, or vandals overturn it. Bees are tough, fortunately, and if you reassemble the hive and get the roof on, they will probably survive. On one occasion I found a hive in a Stoke Gabriel orchard on its side in February, and from the yellowing grass under the detached roof three feet away, it had been like that for at least a fortnight. Thanks to sticky bee-gum (propolis) the six central frames had stayed together, with a tight cluster of bees on four of them. Reassembled and fed lukewarm sugar syrup a few hours later, they came through the rest of the winter and built up into a strong stock by June. Feeding syrup in February is not recommended, but sometimes it just has to be done.

The cluster

It may help if you have a clear picture in your mind of what is going on inside your hive at this time of year. At the beginning of winter, say mid-November, a cold snap will get the bees clustered on the lower half of the six or seven central combs, which will have some empty cells, but be packed solid with food at the top; the outermost four or five combs should be solid with sealed food from top to bottom. During cold weather in December the bees will be in a fairly tight, oval cluster, with the queen in the hollow centre where the temperature will be at least 57–60°F, and probably nearer 70°F. At the outside edge of the cluster the temperature may be only 45°F, and elsewhere in the hive only a degree or two above the average temperature outside. If the cold spell continues for several weeks with the outside temperature never going above 40–42°F even at midday, the cluster will very slowly move upwards as it eats its way through the stored food; when the edge of the cluster reaches the edge of the combs, the bees will flow over and spread on to new ones. Usually, however, there will be a few days of mild weather, and when the outside temperature reaches 46–48°F or more, especially if the sun shines on the hive and warms it up slightly, the cluster will loosen and expand, enabling the bees to move around the hive and resettle on fresh combs still full of food. Some bees will be flying, usually just to clear their bowels, but a few may bring in loads of whitish pollen from that useful evergreen shrub with the clusters of white flowers (laurustinus). When the next cold snap comes along, the cluster will probably be back on the lower half of the combs, but $1\frac{1}{2}$–3 inches to one side of its former position, and before the end of January the temperature in the warm hollow will be up to 95°F, and the first eggs will be laid. With a weak stock this position may not be reached until mid-March.

Do not be unduly worried by the presence of some dead bees in the entrance or on the ground below; your hive population was probably over 25,000 in October and will be down to 10,000 or 12,000 by the middle of January, so the death rate from natural causes must average 150 a day. During most of the year bees die with their boots on away at work, but in winter many die quietly at home, and around noon on a mild winter day you may notice bees on undertaker duties keeping the entrance clear, and even flying the bodies away when weather permits. From mid-February onwards young bees willl be emerging from the sealed brood in increasing numbers, and within a week or two the population will have passed its lowest point and start to increase again, as more and more eggs are laid each day.

A second hive

With just a solitary stock, you can run into problems which could be easily solved if you had a second hive of bees alongside, and even a beginner is urged to aim from the outset at two hives rather than one, even if he never intends to expand beyond this. With this in mind, some time between now and next May you should buy, beg or build a second hive. At some centres there are evening classes at which a hive can be built, but it is quite possible to do this at home. A good woodworker can build from measurements, but the average amateur is strongly recommended to borrow a spare brood box for a few weeks and copy it, using a spare frame to check as you go along that the internal dimensions are right, with the $\frac{1}{4}$-inch bee-space between the frame and the hive walls. Making a floor board, crown board and roof is relatively easy.

During January and February:

a check hives after severe gales;
b start making a second hive;
c note in your new diary what the bees are doing, the time of day and air temperature;
d attend Branch meetings, above all your Branch Annual General Meeting.

EXPERIENCED BEEKEEPERS – JANUARY

It is always said that bees should not be inspected in winter, but much can be learned by inspecting hives from the outside, even at the end of January, according to the following schedule:

a Look carefully on the alighting board and immediately below the entrance on the ground. If all is well there should be a few fine particles of wax, looking almost like dust, arising from a slow, steady uncapping of sealed stores as the bees break open fresh cells to get at the food. Larger pieces of wax, looking like coarse fingernail clippings, would suggest the presence of a mouse, or perhaps a bout of robbing late in autumn by either wasps or other bees. There may be up to a dozen or so dead bees on the ground but at this time of year there should not be more than this. At the same time, check that the entrance is not blocked by dead bees or debris; it is most unlikely that it will be, but bend down and look in just the same.
b Now put your ear against a wall of the brood chamber and listen very carefully. A normal

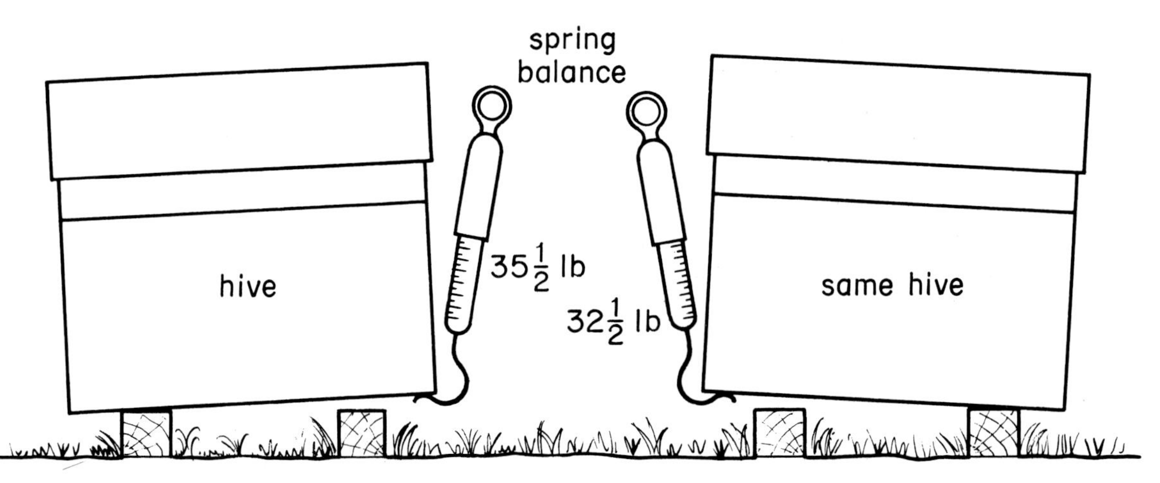

82 Weighing a hive. In this case the hive weight is $35\frac{1}{2}+32\frac{1}{2}=68$ lb – first one side is weighed and then the other.

healthy cluster will not make any sound at all, but a steady, dry, crackling sound (something like the crickle-crackle of bees working on wax foundation in June) would suggest acarine disease, the sound being made by the uneasy movement of infected bees. (See page 172.)

c Now take off the roof, very gently so as not to cause any vibrations which might alert the bees, apply your nose close to the feed hole and sniff. If all is well there will be a warm, wholesome smell vaguely reminiscent of very clean babies and newly baked bread, together with a pleasant smell of wax. If there is any brood disease present you may detect a sour or even offensive odour. If there are no living bees then the smell will be cold, damp and musty. The presence of mice can usually be smelt (not always).

d Gauge the weight of the hive by hefting first one side and then the other. It takes a year or two of experience to detect with certainty an unusually light hive, but by using a spring balance calibrated in 250-g steps (or ½-lb notches) on each side in turn and lifting until the hive just rises, the actual weight may be found by adding the two weights together, and compared with the weight of the hive three months before (middle or end of October). An average stock will have consumed 10–12 lb (about 5 kg) during this three-month period, but if most of the food has gone, the drop in weight may be 25 lb or more.

Action

If the hive is light and has the wrong smell, it should be investigated further, so gently remove crown board and super (if present) and look down through the queen-excluder. If no cluster can be seen, get the excluder off and confirm no live bees present, then remove at least three of the central combs to check for unemerged brood (possibly indicating a brood disease) or bees dead with heads in empty cells (starvation) or some other condition.

If live bees are present but the hive is very light, put a 5-lb block of candy over the feedhole, with an empty super or eke to accommodate it, and replace the roof. If acarine is suspected, then Frow treatment can be applied in February, but this is not always successful and the colony may die out early in spring. Quite often it is a waste of time, and even if they survive they are too weak to build up in time to be honey-gatherers this summer.

The progeny of that queen may be very vulnerable to acarine, and might with advantage be culled, and better use made of the drawn combs (acarine mites do not survive away from live bees).

A practical example written on 28 January 1984 may illustrate these points:

Yesterday 40 hives in five apiaries were checked, the whole operation taking about 1½ hours plus travelling time.

a Roofs secure on all hives; all fitted with mouseguards. No sound audible from any hive. No bees flying. (Air temperature 39°F.) Outside one hive noticed on ground below entrance about 30–40 long-dead bees and layer of wax particles, larger than dust but not large enough to suggest mice. Other hives had either no dead bees at all (was pleasantly surprised by this) or only two or three. Possible that ants, birds, fieldmice may have removed evidence.

b Smell test negative for disease but suggested no live bees present in the one suspect hive, so opened up and checked all combs and floor board. No bees, alive or dead, not even one. Coarse wax fragments on floor, no stored honey but much pollen, no unemerged brood, all cells open. Ragged edges noted on slightly deeper cells in food arch, indicating that they had been uncapped in a hurry by robbing bees (or wasps).

Diagnosis

Probable queen failure in early autumn; in this case fairly sudden total failure, as no evidence of excess drone comb or attempts to re-queen from own eggs. Probability that bees had joined up with another queen-right stock in same apiary, and then robbed out their own hive. Some time ago I watched this actually happening in the home apiary one afternoon in late September, about an hour before I had planned to re-queen the queenless stock. Reference to hive records revealed an entry: 'Fat queen (white marked), brood in all stages, August '83.' So the queen had already brought the colony through two winters (1981 and 1982) and should have been replaced in the autumn of 1983.

Comment

More often than not, a queen will carry on right through her third year, but it is good practice to re-queen all stocks every other year. In any case, most queens do not give of their best after the second year. They are more likely to head a diminishing colony, less likely to produce a good honey crop and more likely to swarm. It should be said here that the useful life of a queen, and of workers too, is not measured by time alone but by work done. A valuable breeder queen, purposely restricted to three or four frames, may continue to lay for three or four years quite satisfactorily.

Sudden and complete queen failure is not typical; often the queen will be laying unfertilized eggs, having run out of sperm, but will still hold the bees together as they dwindle to almost nothing by the time of a normal first inspection at the end of March or early April. In this condition the stock would not be worth re-queening and better use can be made of the drawn combs by putting in an over-wintered nucleus, after routine sterilization of the empty drawn combs with 80% acetic acid.

Acarine testing

This may be simply done in five minutes by the side of a hive, using two pins held close together to impale a bee on its back on a sliced cork held in one hand, and then pushing the head back and off with the small blade of a penknife while looking through a jeweller's magnifying glass (see Fig. 87). It is not possible to see the actual parasites, but the colour and appearance of the large trachea just below the collar can be noted. A milky-white tube shaped something like a clothes hanger indicates that all is well, but a discoloured or brown tube indicates the presence of acarine mites. Government pamphlets obtainable free from the Ministry of Agriculture describe treatment which can be given with folbex strips in the active season, or Frow capsules in November or February.

EMERGENCY FEEDING – FEBRUARY

Candy-feeding

If for any reason the bees were not fed in the autumn and the hives feel light in February, then a large block of candy or baker's fondant (at least 2 kg or 5 lb) placed over the feed hole, close to the bees, is a good insurance. When bees die from starvation they usually do so in March and April; remarkably little food is needed just to keep bees alive in winter, and it is only when the queen starts laying on any scale (early March) that the stores are used up rapidly, with all the babies to feed. By the middle of March it is safe to feed sugar syrup again (although not in a spell of bad weather). Syrup fed in winter excites the bees and they come out to look for nectar at a period when there is none to be found, and either fail to return or suffer wear and tear at a critical period. Candy has about the same water content as the natural sealed stores in the cells, and does not excite the bees.

Candy-making

Can you count up to five? Then you can always remember how to make candy. Just say to yourself: '1, 2, 3, 4, 5 – candy keeps bees alive!' Candy consists of 1 pt of water to 5 lb of sugar, boiled until the thermometer says 234°F. Even if you only have one hive, do not make less than this amount at a time.

83 Emergency feeding: a 1-kg sugar pack is pierced with a hole 1 inch in diameter, cut in the centre, and a quarter-cup of water is poured in. The pack is then placed over the feed hole.

Bring 2 pt water almost to the boil and add about 2 lb sugar while stirring with a long spoon. When this has dissolved and the liquid is coming to the boil again, add sugar a pound at a time until it has all dissolved, stirring continually. Pure water boils at 212°F, and the more sugar that is dissolved, the higher the boiling point. At first the temperature may only be about 228–30°F, so allow it to boil gently for ten minutes or so, until the thermometer indicates 234°F or just over. Cooks and sweet-makers will recognize this as the 'soft-ball stage', when the hot solution dripped on to a cold saucer just forms a soft, solid mass. Take the pan off the stove and stand in a sink of cold water for twenty minutes, or until the temperature drops to 160°F or less, when the mass will begin to show white streaks and go solid around the edge. Now stir vigorously and pour into two ex-icecream plastic boxes (2 l or just under ½-gal size). When cold you will have two 5½-lb blocks of candy, which will keep indefinitely in plastic bags secured with twist-grips. When you need to feed just invert the box over the feed hole, or over the tops of the frames. Do not remove the candy from the box – the bees will do this. Disregard suggestions that candy is 'old-fashioned' and bad for bees because they need water for it. Natural honey usually has 18% or 19% water – slightly less, in fact, than candy made by this recipe. Bees do have to 'invert' the sugar in candy, which causes more work than stores of honey, but they manage it comfortably.

In winter months every hive should have a mouseguard, a well-secured waterproof roof, and top ventilation to allow water vapour to escape. Most important of all, there should be at least 35 lb of food on the combs in October, failing which, feed a 5½-lb block of candy in February and follow up with a gallon of syrup at the end of March.

LIQUEFYING HONEY

Granulated honey

Almost all honey will granulate naturally in 3–6 months, the exceptions being heather, fuchsia and one or two minor sources. Some honeys, notably from oil-seed rape, will granulate in days while still in the comb, and ivy honey will granulate in the comb almost overnight. Sometimes, there is extensive 'frosting' of honey in glass jars, and ill-informed customers sometimes suggest that it is the 'sugar coming out'. In point of fact, the white patches or streaks are partly due to tiny air bubbles coming out of solution on granulation, but mostly due to the dextrose constituent of honey crystallizing in an air space.

Sometimes it is necessary to liquefy this granulated honey for show purposes, or to suit a customer's preference. Obviously honey stored in bulk in 28-lb tins has to be liquefied before it can be bottled.

The problem

Honey is a natural, living food and important constituents such as invertase, diastase and other health-giving enzymes are rapidly destroyed at temperatures over 122°F (50°C). At such temperatures there is also a rapid build-up of HMF (hydroxymethylfurfuraldehyde), a product with a slightly unpleasant flavour which is alleged to be harmful. At temperatures above 150°F (65°C) honey starts to darken in colour and acquires a 'cooked' taste due to partial caramelization. So, if honey is to be kept in good condition, high temperatures (even of water well below boiling point) have to be avoided.

A 28-lb tin holds a cylinder of solid honey 10½ inches tall by 8½ inches diameter, so that heat has to penetrate a considerable distance in order to liquefy the honey throughout. If too much heat be applied too quickly, the honey on the surface will be far too hot before enough heat can reach the honey in the centre to melt it.

A simple solution

If one is concerned with only a few jars, the

84 Honey-warming cabinet, involving the use of a 60-watt bulb.

easiest solution is probably to stand them for a few days in the airing cupboard next to the hot-water cylinder, but in these fuel-conscious days the cylinder may be so well lagged that the honey would not get warm enough. Standing the jars in boiling water is not recommended.

On the principle that 'what keeps heat out should also keep heat in', I have for many years used an old fridge with the freezing box and inside fittings stripped out, leaving just one strong wire rack, capable of being slotted in either high up in the cabinet to take two dozen 1-lb jars, or low down to support two 28-lb tins. With a 40-watt bulb lying on the fridge floor, the temperature inside will build up to 108–110°F when the surrounding air is around 50–60°F. Experience shows that two dozen 1-lb jars will completely liquefy in 12 hours, or two full 28-lb tins in about 36–48 hours. If left a little longer than this, no great harm is done, as at about 110°F the heat loss by conduction through the fridge walls just about keeps pace with the modest heat input of the light bulb, and a steady state is reached. During winter it may be necessary to use a 60-watt bulb, depending on the surrounding air temperature.

One bonus is that honey warmed to this temperature flows very easily through a fine filter into the bottling tank. In fact, because of this, my usual practice is to bottle only immediate requirements at harvest time, all other honey going straight from the extractor into 28-lb tins for bulk storage.

85 Filtering warm honey.

Precautions

With a light bulb immediately below a 28-lb tin there would be a local 'hot spot' where honey temperature could build up, so it is best to have a heat spreader made of a 16-in × 4-in piece of stout aluminium foil or thin tin-plate, bent to make a 'bridge' 4 in × 8 in over the bulb. If a bulb stronger than 60 watt be used, it is possible for honey temperatures to exceed the desired maximum of 112–15°F, and the situation must be watched; switching off for 3–4 hours to allow the heat to 'soak in' can help. If plastic 30-lb pails (ex-baker's or confectioner's) are used, it is even more important to avoid a 'hot spot', and a thin sheet of tin-plate or aluminium should also be placed on the rack below the bucket.

Used with a 100-watt bulb, the old fridge becomes a 'slow oven' ideal for meringues or for drying apple rings or split plums. No doubt our ingenious reader will find still more uses!

Creamed honey

If a 28-lb tin of honey is warmed for a shorter time, there will be an outer layer of melted honey surrounding an inner mass of warm, soft but still fairly solid honey. This mixture may be 'creamed' by using a metal plunging disc with holes (sold by bee appliance dealers), pushed up and down vigorously for several minutes until the contents of the tin are uniformly like very thick cream. (See Fig. 86.) This may be poured and ladled into jars for sale, and is much appreciated by many customers.

HEALTHY HONEYBEES

Those of us who organize or attend talks on the diseases of bees soon become aware from low attendances that this is the least popular aspect of beekeeping. This book attempts to cover the subject in a positive way, but for ease of

86 Creaming honey.

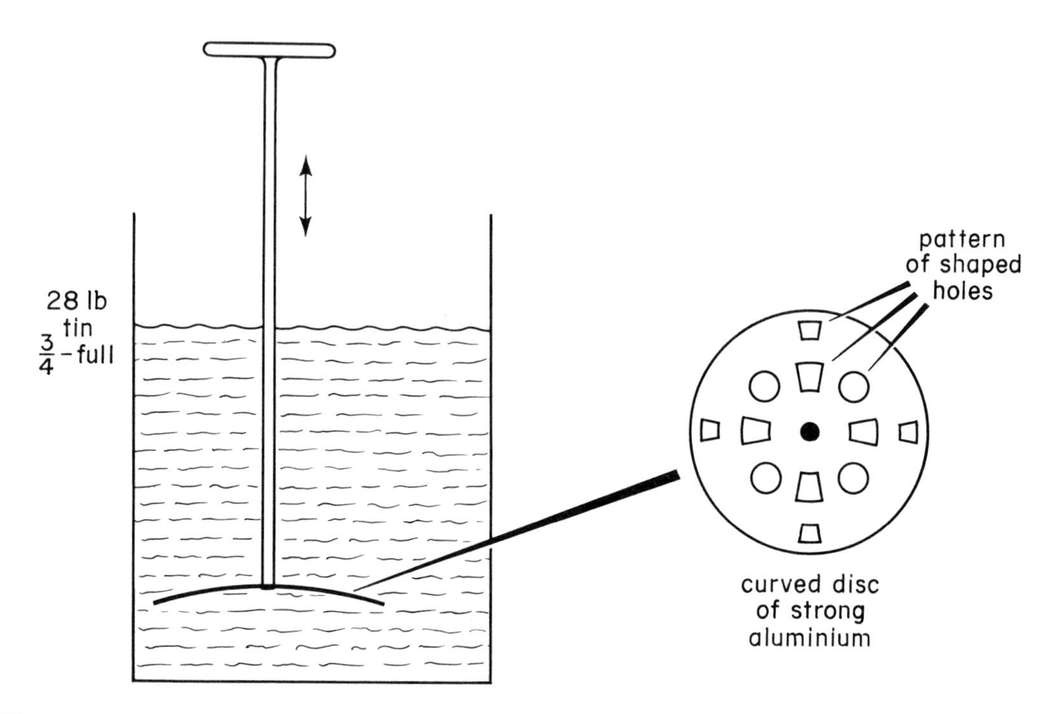

reference lists the main symptoms of 'unusual conditions' and suggests the best remedial action.

The most important single factor is sensible management, making sure that hives are sound and waterproof, fitted with mouseproof entrance in winter, restricted entrance in wasp-time (August to October) and plenty of food at all times. No queen should be older than two years, on good brood combs of which two or three are culled out each year and replaced with wax foundation in early summer. Top ventilation in winter is recommended.

Be content with only two or three hives in a poor area. In a good area where up to 20 hives can be kept on one site, see that the hives are arranged in irregular groups to cut down 'drifting', whereby foraging bees accidentally enter (and when carrying food are accepted into) hives other than their own, spreading disease should there be any.

The problems are described here with the most common first and the least common last.

Unusual conditions

Starvation

Kills more colonies than any disease, and is much easier to deal with. Generous feeding in the autumn and a watchful eye in the spring will effectively prevent this. Stocks well fed in the autumn but with unprotected entrances may host an unwelcome family of mice, resulting in starvation, or at least in stressed bees, loss of combs and a filthy hive.

A wide-open entrance from August to October can result in robbing by wasps (or strange bees), so diminishing food stocks as to cause starvation by spring.

Queen failure

This may occur during winter when the bees can do nothing about it. Usually blamed on an elderly queen, but can happen to a younger queen incompletely mated in a poor summer, so that although alive she becomes a drone-layer. The condition can be dealt with in March or April by removing the queen (if present) and uniting with an overwintered nucleus having a young queen. If there is not a complete breakdown but just a 'slow queen', i.e. abnormally slow spring build-up for no apparent reason, with perhaps only two frames of brood when all other hives have at least four to seven, then remove the old queen and introduce a young one, from an overwintered nucleus.

Nosema

As endemic to bees as the common cold is to humans. As a bee-microscopist would put it, 'If you look hard enough you will always find one or two rice grains (nosema spores).' In one sense it is a 'stress disease', affecting colonies which are moved or over-manipulated, or sited in an area providing insufficient pollen or nectar. In a winter or spring following a bad summer it may be more evident. Bees that have been weakened by acarine or some virus disease like paralysis, allowed to run down under a poor queen or just with insufficient food, are more likely to suffer. As Shakespeare said, 'When troubles come, they come not single spies but in battalions.' We ourselves react similarly, and are more susceptible to illness when our vitality is at a low ebb. Just as in recent years we have become more aware of the importance of sensible diet and lifestyle, as opposed to treatment with more and more medicines, so it is with bees.

To be more specific, nosema is caused by a minute organism which develops and multiplies in the gut of adult bees; the larvae are not affected at all. The effect on bees is to weaken rather than to kill. They live only half as long, their wax glands shrink, their brood-food glands are undeveloped, they are less efficient at nectar collection. In winter infected bees have less protein in their fat layers, and their bowels are distended with sloughed-off epithelial cells and nosema spores, so that symptoms of dysentery can be seen on soiled combs. In spring the colony either dwindles, or expands only very slowly, but may still recover for a while and seem almost normal by summer when bees fly freely and defecate outside the hive. Such a colony usually develops too late to store much honey, and the infection lives on in the soiled combs.

Fortunately there is a specific remedy: an antibiotic called Fumidil B, which together with routine sterilization of brood combs not in use with 80% acetic acid, and blow-lamping floors and brood boxes before re-use, can overcome the problem. These techniques are described in detail on page 26, but are no substitute for good management.

The actual mechanism of infection or re-infection is usually from nosema spores in faeces, either on soiled brood combs or hive parts. It also occurs when bees are crushed and other workers clean up the sticky mess with their tongues, so that millions of nosema spores contained in the liquid mass of crushed bee stomachs and honey are absorbed by other bees, perhaps 50 being infected directly from one crushed bee. Hence the motto, 'Never crush a bee.'

Acarine

Acarine or Isle of Wight disease is caused by a small mite which develops and lives in the large tracheae or breathing tubes. These mites pierce the trachea walls and suck the blood of the bee. Although still able to fly in most cases, efficiency is reduced, and the bees behave like aeroplanes with only three or maybe two engines working instead of four. They may be seen trying to climb up grass stems and clustering in little knots unable to develop enough power for lift-off. Sometimes they will be seen walking away from the hive, with wings at an angle giving the impression of a double 'K', probably because the wing roots are close to the site of infestation in the thorax. In early spring an infected colony is restless and an ear pressed against the outside of the brood chamber can detect a crackling, brittle noise caused by restless bees. It is usually said that the trouble can be confirmed only by use of a microscope, but a simpler method of diagnosis has been described on page 166. There are specific remedies, but in practice the best solution is to keep only bees which are naturally resistant to this disease. Queens imported from areas free from acarine (the USA, for example) may be excellent in other respects, but their progeny is vulnerable to acarine. It seems that genetically it may depend on the time when the hairs normally covering the main thoracic spiracle (breathing hole) harden up. The mite can enter a young bee via this hole only during the first four days of the bee's life, and finds this more difficult with bees whose body hairs stiffen a day or two earlier and act as a barrier. Infection can only take place from bee to bee, and the mite cannot live more than a day away from live bees. Thus when large numbers of bees are kept on one site, especially if hives are arranged in straight lines, the trouble can be spread by bees drifting, or accidentally entering the wrong hive.

Some commercial beekeepers find that oil of wintergreen (methyl salicylate) helps to prevent acarine. They use a felt pad in a small flat tin with holes, or a very small bottle with a protruding wick, placed on the hive floor. Dosage is a teaspoonful (4 ml) given in August and renewed in March.

Mouldy combs and chalky pollen

This can be quite widespread and a great nuisance, but is not serious. A strong colony will not usually have any problems of this kind, but when the bees are covering only two or three combs in early spring, the flank combs will often be affected by damp from condensation and subsequent mould. Good through-ventilation can prevent this, and it helps if the two outermost combs of food (one on each side) are removed in later autumn and given back in March. Any pollen not covered with honey and capped will usually go mouldy, affected by the fungus *Bettsia alvei*, unless it is in the area actually covered by the bees. Pollen is sometimes stored above the excluder in the lower part of the central honeycombs. It does not spin out with the honey when centrifuged and if left will certainly be hard and chalky by next year, giving the bees unnecessary work in its removal. As explained on page 31, it should be scraped out, or the combs placed centrally in the super left on the hive as a food reservoir.

Chalk brood

This is fairly widespread but not usually serious. In early summer a small number of hard, white, mummified larvae may be seen, especially near the edge of the brood nest area. The problem can be distinguished from

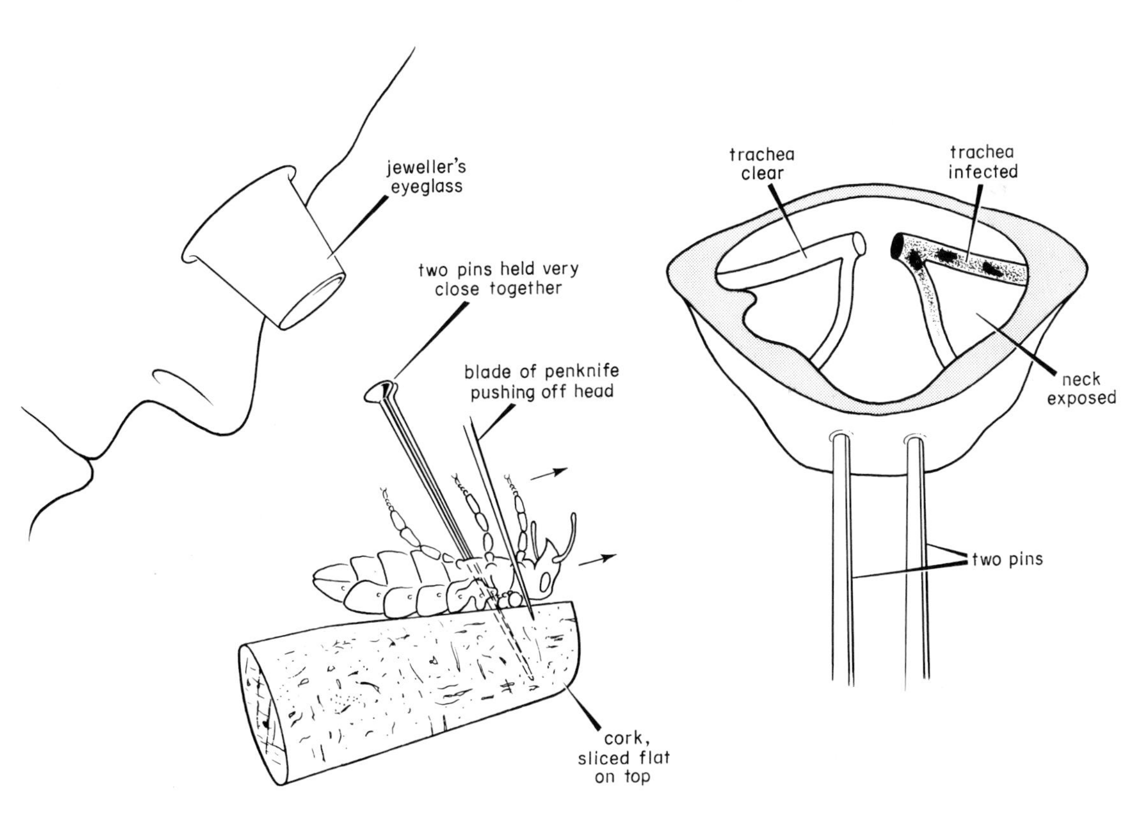

87 Testing for acarine in the field.

chalky, mouldy pollen by probing, when pollen crumbles but chalk brood does not. Caused by the fungus *Ascosphaera apis*, probably present in most colonies but only noticed when the colony is weakened by some other condition, or when there are insufficient bees to cover larvae, especially just before sealing. Keep brood chamber warm in March and April, and re-queen if at all serious, to get progeny with higher natural resistance.

Laying workers

This condition arises only when a colony has been queenless for a long time in summer, oddly enough sooner in a stock with fairly new brood combs than when old, much-used brood combs are present. A number of workers will have their normally immature ovaries developed to a point when they can each lay a few eggs a day, and as these are not fertile they produce drones, but smaller than usual. Characteristically one finds cells with more than one egg, eggs deposited on the sides of cells rather than the base, but above all patches of high-domed drone brood scattered all over the brood nest. The rest of the workers will all be fairly old but may be numerous enough to be worth saving. There will be large numbers of undersized drones on the combs and flying from the hive. If the population is small, the colony is probably not worth saving, and the best solution is to take out each comb and shake the bees off in the centre of the apiary, when some at least will join other colonies and do useful work. If one takes a swarm, then it can be shaken directly into the laying worker colony, where it will take over by sheer pressure of numbers.

Recently I discussed this problem with M. Jean Scrive, President, Queen Rearers' Association of France. We agreed that no normal method of queen introduction was successful with a layer worker colony. He told me of his new method, used successfully nine times in the last two years. He dips a queen in royal jelly and drops her directly into the hive. Any amateur with half a teaspoon of royal jelly collected from a few swarm cells can do the same.

Dysentery

Sometimes in spring one finds the front of a hive soiled by streaks of bee excreta. This does not necessarily mean that disease is present, especially if the symptom follows a long spell of cold weather when the bees have been unable to come out on cleansing flights. Dysentery is sometimes more noticeable in hives which were taken to the heather the previous August and not fed syrup on return. The high protein content of heather honey can cause bowel congestion when it is too cold for flight over a period of several weeks. If the actual combs and frames inside the hive are soiled, then the dysentery is probably associated with nosema, but may have been caused by unripe honey, or by honey granulated in the combs. A honey with a glucose content higher than normal (oil-seed rape, ivy) may granulate in the combs, and in so doing the glucose crystals separate out leaving a fructose solution more watery than usual around the actual crystal, and this is sucked up by the bees. The extra water builds up in the lower abdomen in winter and in the absence of indoor toilet facilities gets deposited on frames and comb surfaces. Any nosema present will leave spores in the excreta and infect healthy bees. The condition should be noted and samples of bees sent for examination.

Paralysis

Paralysis in one form or another, caused by a virus, can weaken a colony to the point where no honey surplus will be obtained. The affected bees may be dark and shiny, with no hair. They may be seen to shake and nibble at each other as if irritated. To date there is no cure for this condition except to re-queen and hope that the progeny of the new queen will have greater natural resistance.

Other minor abnormalities

Sometimes one can notice a few bees with small reddish-brown 'bee fleas' on their backs; these are tiny flightless insects called braula, which do not damage the bees and are not parasites, in the sense that they do not bite and suck from the body fluid, but merely help themselves to food. The queen is more likely to be infested than most of her workers, but this condition is seldom serious, except that in the larval stage, braula coeca tunnel under the wax cappings and can make honey in the comb unsaleable by disfiguring the surface. Tobacco smoke will dislodge the creatures, but usually nothing need be done. Should the condition be more general, it is recommended that a pea-sized crystal of PDB (paradichlorobenzene) be placed on a sheet of stiff paper and pushed in the entrance, or alternatively some cigarette ends rolled in corrugated packing material and put in the smoker. Before smoking, push a sheet of stiff paper on to the hive floor to catch the braula as they fall off, stupefied by the tobacco smoke, and remove the paper an hour later before they recover and get back on to the bees. Minor brood diseases (chalk brood, sac brood, addled brood) may be noted from time to time, but these conditions usually right themselves and seldom cause serious trouble.

Wax moth

Wax moth may be found in almost any colony at some time or other, and the larvae can cause serious damage, and can even destroy combs not covered by bees. The section on beeswax covers this (see page 128).

Spray poisoning

Spray poisoning can be extremely serious, and should be suspected when large numbers of dead bees are found in front of a hive in summer, with perhaps some bees spinning and turning around on the ground in obvious distress. Diagnosis should be confirmed by sending several hundred dead bees to the Ministry of Agriculture at Luddington for analysis *immediately*, packed in a ventilated box, not an airtight plastic container, together with a report of the circumstances and, if possible, details of crops grown locally which might have been sprayed. There are schemes for liaison between farmers and beekeepers, but for various reasons they often fail to prevent the trouble. It is to be hoped that the problem will eventually abate, as newer formulations are being evolved which can be used with less damage to bees. Even if a beekeeper is notified in advance, the actual spraying may take place days later than expected, as the spray contractor may be overbooked, and in any case it is not always practicable to move hives at short notice or restrict the flight of bees. Large-scale beekeepers on pollination contracts are likely to be in touch with possible spraying programmes. All that the amateur with one or two hives living in the danger area can do, is to keep in touch and perhaps put a couple of forkfuls of hay or straw loosely over the hive entrance. This seems to occupy the bees and check foraging for the day without causing any harm.

American foul brood

The really serious brood conditions have been left until last, partly because they are dealt with professionally by the Bees' Officer (usually referred to as the 'Foul Brood Officer') and partly because they are less common hive conditions than those already covered.

American foul brood affects larvae, which collapse and die *after* cells have been sealed; their colour changes to a creamy brown (ropy) and finally to a very dark brown, almost black scale lying on the lower side of the cell from just behind the cell mouth right back to the base. The cell cappings become moist and darker, and collapse inwards as the larva shrinks. In their attempts to remove dead larvae the bees may nibble holes in the cappings and sometimes chew them away completely. The organism is *Bacillus larvae*, rod-shaped, $2\frac{1}{2}$–5 microns long by 0.5–0.8 microns wide. It is so menacing because it forms countless millions of oval spores (1.3×0.6 microns) in each dead larva, and these spores are highly resistant to drying, heat and chemical disinfection. They can remain viable for many years. The actual infection is via the spores, which germinate in the mid-gut of very young larvae; larvae more than two days old very seldom become infected. The disease is usually spread by a strong stock robbing out a weak, infected stock, by nuclei made up from infected stock, by infected honey extracted from supers or left in extractors and tins for bees to clean out, and by young bees drifting into other hives after midday 'play flights'.

Where the disease has been suppressed by antibiotics (usually terramycin), hives may still yield a surplus, but remain reservoirs of infection which will flare up if not given terramycin. The only really effective remedy is destruction by fire. So far as the beekeeper is concerned, any hive showing symptoms of AFB must be reported immediately to the local Bees' Officer, who will check and burn, after confirmation by microscopy. The effectiveness of this drastic treatment was well illustrated in Wiltshire recently, when a team of Bees' Officers in 1980 cleaned up some 240 cases of AFB. In the following season there was no evidence that it had spread and in 1983 only 13 cases were discovered in the whole of that county.

European foul brood

This also affects larvae, but death occurs *before* cells have been sealed. The actual cause of death is starvation, as the organism competes for food in the larval gut. The larvae lie uneasy in their cells at unusual angles, and die when about four or five days old. After death they turn brown and decompose; a sour smell may be noticed. The organism is *Mellissa coccus pluton*, but after death *Bacillus alvei* usually takes over and finally an almost pure culture of *Bacillus alvei* remains.

The ebb and flow of this disease is not easy to understand, as colonies may appear to grow out of it during the summer while still remaining infected. At the peak of a honey flow the proportion of nurse bees is less, so less larval food is available and a higher proportion of larvae shrink and are ejected by the bees. When bad weather interrupts a nectar flow, the queen lays fewer eggs and a temporary excess of nurse bees is able to provide more food, enough for larvae and their cocci, so that larvae live on and are not ejected. When the main honey flow starts up, brood-rearing increases again, and many infected larvae are deprived of surplus glandular food and die. No highly resistant spores are formed, and dried-up cocci survive no longer than three or four years. Infection is by contact with other hives in areas where the disease is endemic, from faecal remains in cells or combs. Any beekeeper aware or believing that a stock is infected must contact the local Bees' Officer, who will check. In Britain treatment by terramycin *administered by the Bees' Officer* is permissible, but often the recommendation is to burn. In most countries terramycin is freely available and extensively used. Where facilities exist, the equipment can be fumigated with ethylene oxide. At the 1981 Quebec Symposium on EFB, experiments using lactic acid were reported as a control measure, but terramycin is still the standard alternative to destruction.

Varroa

The mite itself is crab-shaped and of pinhead size. Varroa has lived as a parasite for thousands of years on the smaller Cerana bee of the Far East, and has evolved a system of survival without actually killing its host. The mechanism evolved was its life cycle, capable of taking place only in drone brood. Thus a percentage of drones of which all bee colonies produce a surplus, was the tax levied on Cerana bee colonies. However, honey bees in the UK have a longer life cycle (21 days for workers instead of Cerana's 18–19 days), so that while still preferring honey bee drone cells (by a factor of 7:1), this pest can also breed in worker cells and hence is lethal rather than just a nuisance. It can also breed much faster in our mellifera drone cells (24 day cycle) than in those of Cerana.

In the early stages of infestation, only one female varroa enters a brood cell, feeding heavily on the larva for 2 days and then laying up to 6 large eggs at 30 hour intervals. One of the first two eggs produces a male and the rest females, which mate (if time allows) with their brother, who then dies in the cell. Only mated females leave the cell. About 25% of females repeat this cycle once, and 5% twice, living on adult bees for a few days in between.

Effects on bees

During the summer breeding season, 85% of the varroa population live inside bee brood cells, inflicting multiple injuries on both larvae and pupae. Even if only one varroa female has entered a brood cell, the bee will emerge seriously deformed with a very short life, possibly unable to fly. In the initial stages, when relatively few mites are present in a hive, most of the victims are drones and the effect on the colony appears not to be serious. As the mite population builds up, there are not enough drone cells available for them, and females then breed in worker cells; the colony declines

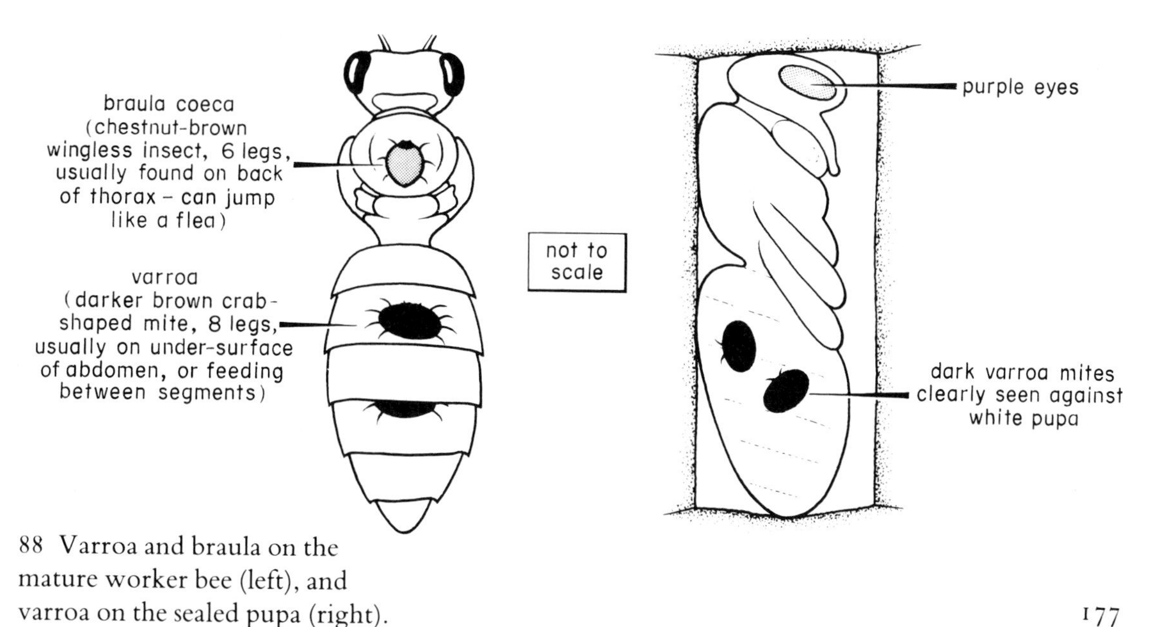

88 Varroa and braula on the mature worker bee (left), and varroa on the sealed pupa (right).

rapidly and gradually dies out. Queen larvae are rarely infested, as queens develop from egg to emerged insect in only 15 days, as opposed to 21 days for workers and 24 for drones.

The final collapse of a colony of bees at the end of perhaps the third summer of infestation is relatively sudden, as if stronger varroa were being bred. This may be linked with hybrid vigour due to cross-mating, when several female mites are laying in one brood cell and brother/sister mating gives way to cross mating.

Spread of varroa

Varroa itself spreads only slowly, by bees drifting from one hive to another, by strong colonies robbing out weaker, infested colonies and by swarms as well as by individual drones. Mankind brought varroa from the Far East to Europe by moving hives, and to South America via escort bees in queen cages. Within newly infested regions mites were spread by moving hives for pollination, as in the USA, as well as by moving combs from hive to hive.

Detection

When there is no brood, the mites survive only on live bees, scarcely feeding and transferring to other bees if their host dies.

a A crude but very effective diagnosis may be obtained by dropping a large handful of bees into a honey jar half-full of petrol and swirling it around. When poured first through a coarse wire filter (catching the bees) and then through a filter paper, the mites may be counted on the white paper. This method would be effective in assessing the degree of infestation of a swarm.

b At any routine hive inspection in summer, drop a sheet of light coloured paper smeared with Vaseline around the edges into an empty nuc box. Select two frames of brood well covered with bees and place in the nuc box. Puff smoke over and between the two frames and cover with a loose sack for 10–15 minutes. Then replace the two frames and check the paper for mites.

c From April onwards there will usually be drone brood present. Uncap these cells with a wide uncapping fork, when any mites present will show up clearly as brown or black spots on a white background.

d When changing floors in spring, carefully scrape the debris into a cardboard container and sent to M.A.F.F. for a routine check, as recommended by the B.B.K.A. Also have a look yourself with a magnifying glass – the mites are not difficult to see.

Treatment

a Beekeepers on the continent are currently obtaining good results with 2 hessian strips impregnated with fluvalinate (apistan), inserted between combs in the brood chamber in September and again in February. Fluvalinate, $C_{26}H_{22}CF_3N_2O_3$, is a contact, acaricide not dangerous to bees, or to humans (except as a mild eye irritant). The human lethal dose is over 1.4 Kg. Having a very low vapour pressure (B.Pt. over 450°C) it is very persistent and tends to be retained on wax combs, so that it remains effective for a long time, well after the month that the strips are usually left in position, but with no known harmful effects.

b German beekeepers are also having success with perizin dripped on to frames, and with formic acid.

c There are also several management techniques in use, e.g.

(i) Insertion of a composite drone comb at the edge of the brood nest early in April and its removal 3½ weeks later to destroy drone brood plus varroa attracted to it.

(ii) Caging a comb plus queen at the end of June for 3–4 weeks, with worker access via QX strips, so that mites can migrate

89 Degree of infestation

A reasonable assessment of the degree of infestation may be obtained by placing a clean sheet of paper lightly rubbed over with Vaseline on the hive floor and withdrawing it 24 hours later. Beekeepers in Germany have linked the normal daily mortality of varroa mites as counted, with the degree of infestation as shown in the table below. (Acknowledgment to Prof. W. Ritter)

Grade of Infestation	No. of dead mites per day	Infestation on drone comb	Infestation on worker comb	Infestation on adult bees	Observed colony behaviour
Light	0–4	Occasional	Not noticed	Not noticed	Normal
Medium	5–10	Clearly visible	Seldom noticed	Not noticed	Normal
Heavy	11–15	Almost everywhere	Clearly visible	Sometimes noticed	Fairly normal
Critical	16+	Complete	Widespread, looks like foul brood	Crippled, stunted bees seen	Irritated, bees no longer clustered on brood.

to the only brood available, after which the comb is hung up for birds to clean out and the queen run in onto another frame, or replaced. Trials in Devon for the last three years (no varroa) have indicated the bonus of more honey in the supers, with no problems.

(iii) Working on the principle that mites are killed at 46°C/115°F while bees are only stressed temporarily, beekeepers in Japan and Russia are using carefully controlled thermal chambers plus frames of bees, from which the mites drop off.

Research

It is widely appreciated that mites could become immune to any one chemical treatment; also that a build-up of chemical residues could cause problems, especially when more than one chemical at a time is involved. Thus a vigorous campaign of research is being conducted in different centres, for example:

a At Oberursel in Germany research workers are cross-mating their Carnicans with Capensis bees from South Africa, with the object of obtaining a hybrid with a shorter brood life cycle.

b At Bures-sur-Yvette in France some basic research is in progress to identify any pheromone by which varroa mites are more strongly attracted to drone brood (by a factor of 7:1). Obviously any such pheromone identified and synthesized could be used in a bee-proof varroa trap, with no chemicals in contact with bees or combs.

c It was observed that when an infested colony swarmed, the great majority of mites were left behind in the swarmed stock. Experimental work quantifying this effect has given some interesting results, e.g.

30000 +	1200 → swarm of	15000 + stk	15000
bees	mites	bees	bees
		200 mites	1000 mites
30000 +	50 → swarm of	15000 + stk	15000
bees	mites	bees	bees
		NO MITES	50 mites

This suggests the possibility of mite control by artificial swarming, even a re-appraisal of our current attitude to the collection of swarms!

It remains to be seen whether the English Channel will continue to be as effective a barrier against varroa as it was against Napoleon and Hitler. In the meantime no beekeeper should illicitly import bees or queens from the Continent or America, and pressure must be kept up to maintain strict control at ports lest a swarm or cast enter Britain via one of those 'long vehicles' coming over on the ferries, or on the mast of a ship.

Appendix A
Bees and the law

In the absence of any statute dealing specifically with the keeping of bees and title to swarms, such matters are dealt with under common law, based on judgements already given in previous cases. The basic concept involved is that a beekeeper (like any other citizen) has a duty to behave as a 'reasonable person', and a person who keeps bees of such a nature, or in such numbers, or in such a position, as to deny a neighbour the normal and reasonable enjoyment of his house, garden or property, is regarded as committing a nuisance. Contravention of the legal principle *sic utere tuo ut non alienum laedas* might be termed 'negligent or inconsiderate disregard of the duty of one neighbour towards another'. If this principle of *sic utere* can be invoked against a beekeeper whose bees annoy or sting a neighbour, one might suppose that a beekeeper could equally take action against a neighbour who sprayed his fruit trees with a poisonous substance *while they were in blossom*. The Protection of Animals Act, 1911 (as amended 1927), covers 'any animal ... sufficiently tamed to serve some purpose for the use of man'.

So far as swarms are concerned, the position seems not to have changed since Justinian and the days of ancient Rome. Bees are regarded as *ferae naturae*, i.e. as the property of a beekeeper only until they break away from his custody, when they become the property of their next captor. Their natural liberty is deemed to have been recovered when they escape from the sight of the beekeeper or, though continuing in his sight, when they are so placed as to be difficult to recapture. A beekeeper has no basic right to enter a neighbour's property without permission to take a swarm, and risks an action for trespass should he do so.

Just as a dog is commonly said to be 'allowed one free bite' before being regarded as vicious, so an isolated case of stinging or swarming would not be held to constitute a nuisance. Repeated stinging or swarming, established to have arisen from a neighbour's bees, would constitute a course of conduct rightly considered a nuisance, the more readily so if it could be shown to be due to negligence on the part of the beekeeper.

BEE DISEASE LEGISLATION

The Bees Act, 1980

(An 'enabling' act.) Gives power to MAFF to make orders to prevent the introduction and spread within Great Britain of pests and diseases affecting bees. Any contravention of such orders may lead to a fine of up to £1,000.

The Importation of Bees Order, 1980
Prohibits the import of all bees into Great Britain except under an import licence. Such a licence is issued for a particular consignment or consignments which must be accompanied by a health certificate. In respect of queen bees, attendant workers must be replaced and sent to the Ministry of Agriculture at Luddington for examination; the importer must also keep a record of the ultimate destination of each queen.

The Bee Diseases Control Order, 1982
Imposes on any beekeeper who has reasonable grounds to suppose that his bees may be infected with American foul brood, European foul brood or varroasis the obligation: *a* to report it; *b* not to remove any bees, bee products or equipment until so authorized by the proper authority. In his official capacity an authorized person (normally the Bees' Officer, commonly called the 'Foul Brood Officer') may take samples, mark any hive or equipment, and issue a 'standstill order' when he has reasonable grounds for suspecting disease, or is refused entry to a place where he believes bees, equipment or produce are present.

AFB
Should disease be confirmed by laboratory tests, a notice will be served requiring destruction by fire of all bees, combs, honey and quilts, and at the option of the beekeeper either to destroy by fire, or treat in such a manner and with such substances as may be specified in the notice, all equipment which appears to the authorized person to be infected or to have been exposed to infection. Any form of destruction or treatment must be done in the presence of the authorized person. Where a beekeeper agrees by signing the order, an authorized person may serve a destruction order without waiting for a laboratory test.

EFB
If disease is confirmed, a very similar notice will be served, except that there is the option of treatment as an alternative to destruction. Where bees are treated they must remain in the custody of the beekeeper for eight weeks and no combs or product removed during that period except under licence.

Varroasis
An order may be served requiring the destruction, treatment or isolation of the hive, bees, combs, quilts, products and appliances which may have been affected. In addition an area may be declared as infected, within which an order may be placed requiring the destruction, treatment or isolation as stated above. Any movement into, within or out of the area will be prohibited. In all cases the notice will give details of methods of destruction or treatment and timing, and whether to be carried out in the presence of or by the authorized person.

SALE AND SUPPLY OF HONEY PRODUCED IN THE UNITED KINGDOM

This is subject to the following acts and regulations:

Food and Drug Act, 1955
The Labelling of Food Regulations, 1970, as amended
The Honey Regulations, 1976
The Materials and Articles in Contact with Food Regulations, 1978
Weights and Measures Acts, 1963 to 1979
The Weights and Measures (Marking of Goods and Abbreviations of Units) Regulations, 1975, as amended
The Weights and Measures Act, 1963, (Honey) Order, 1976
Trade Descriptions Acts, 1968 and 1972

Appendix A – BEES AND THE LAW

The Trade Descriptions (Indication of Origin) (Exemption No. 1) Directions, 1972
Consumer Safety Act, 1978
The Glazed Ceramic Ware (Safety) Regulations, 1975

Legal definitions

'Honey' means the fluid, viscous or crystallized food which is produced by honeybees from the nectar of blossoms, or from secretions of, or found on, living parts of plants other than blossoms, which honeybees collect, transform, combine with substances of their own and store and leave to mature in honeycombs.

'Comb honey' means honey stored by honeybees in the cells of freshly built broodless combs and intended to be sold in sealed whole combs or in parts of such combs.

'Chunk honey' means honey which contains at least one piece of comb honey.

'Blossom honey' means honey produced wholly or mainly from the nectar of blossoms.

'Honeydew honey' means honey, the colour of which is light brown, greenish brown, black or any intermediate colour, produced wholly or mainly from secretions of or found on living parts of plants other than blossoms.

'Drained honey' means honey obtained by draining uncapped broodless honeycombs.

'Extracted honey' means honey obtained by centrifuging uncapped broodless honeycombs.

'Pressed honey' means honey obtained by pressing broodless honeycombs with or without the application of moderate heat.

Markings on containers

Honey should be prepacked for retail sale or otherwise made up in a container for sale only if the container is marked with the following information:

1 An indication of quantity by net weight in both imperial and metric units.
2 The name or trade name and address of the producer, packer or seller.
3 A description in one of the following forms:
   a honey;
   b comb honey;
   c chunk honey;
   d baker's honey or industrial honey;
   e the word 'honey' with a regional, topographical or territorial reference, e.g. Devon honey, honey from South Devon, moorland honey;
   f the word 'honey' with a reference to the blossom or plant origin, e.g. heather honey, lime honey;
   g The word 'honey' with any other true description, e.g. honeydew honey, pressed honey, set honey.

In wholesale transactions of containers of a net weight of 10 kg or more, a separate document showing the required information is sufficient if it accompanies the container.

Composition of honey

1 There should be no addition of substances other than honey.
2 The honey should as far as practicable be free from mould, insects, insect debris, brood and any other organic or inorganic substance foreign to the composition of honey. Honey with these defects should not be used as an ingredient of any other food.
3 The acidity should not be artificially changed and there is a legal maximum level of acidity.
4 Any honeydew honey or blend of any honeydew honey with blossom honey should have an apparent reducing sugar (invert sugar) content of not less than 60% and an apparent sucrose content of not more than 10%.
5 Honey with a moisture content of more than 25% should not be supplied.

Baker's or industrial honey

Honeys of the following descriptions should be labelled or documented only as 'baker's honey' or 'industrial honey'.

1 Heather honey or clover honey with a moisture content of more than 23%.
2 Other honey with a moisture content of more than 21%.
3 Honey with any foreign taste or odour.
4 Honey which has begun to ferment or effervesce.
5 Honey which has been heated to such an extent that its natural enzymes have been destroyed or made inactive.
6 Honey with a diastase activity of less than 4 or, if it has a naturally low enzyme content, less than 3.
7 Honey with an hydroxymethylfurfural (HMF) content of more than 80 mg per kg.

Misdescription

There are two basic types of illegal misdescription; the direct and the indirect or misleading.

The direct misdescription should be obvious and can be fraudulent. A simple example would be to describe Australian honey as 'Devon honey'.

Careful thought will avoid indirect misdescription. Examples of such misdescriptions could be as follows:

a An illustration of bees collecting nectar in a moorland setting on honey which is not from moorland.
b The statement 'Produced in Devon' applied to honey which is blended in Devon from honeys of various origins which may or may not include Devon.

The following guidelines should be observed:

1 Any reference, direct or indirect, in words or by means of any pictorial device to the blossom or plant origin should only be applied to honey derived wholly or mainly from the blossom or plant indicated.
2 Any reference to the regional, topographical or territorial origin of the honey should only be applied to honey which originated *wholly* in the region, place or territory indicated.
3 Honey produced outside the United Kingdom which has a United Kingdom name or mark should be accompanied by a conspicuous indication of the country in which the honey was produced. Blends of honeys from two or more countries, which may include the United Kingdom, may be accompanied instead by a *conspicuous* indication that it was produced in more than one country.

Appendix B
Useful addresses

Publications (in Britain)

BBKA News, David Charles, 'Bickerton', Church Lane, West Pennard, Somerset.

Bee Craft, Mrs White, 15 West Way, Copthorne Bank, Crawley, Sussex.

Beekeepers' Annual, John Phipps, Walnut Cottage, Pilham, Gainsborough, Lincs.

Beekeeping, P. P. Rosenfelt, 42a Clifford Street, Chudleigh, Devon.

British Bee Journal, Cecil Tonsley, 46 Queen Street, Geddington, Kettering, Northants.

Irish Beekeeper, James Doran, St Judes, Mooncoin, Waterford, Eire.

Home Farm, Katie Thear, Broad Leys Publishing Company, Widdington, Saffron Walden, Essex.

Scottish Beekeeper, N. Blair, 44 Dalhousie Road, Kilbarchan, Renfrewshire.

MAFF Publications, Tolcarne Drive, Pinner, Middlesex.

Publications (abroad)

American Bee Journal, Hamilton, Ill 63241, USA.

Australian Bee Journal, PO Box 426, Benalla 3672, Victoria.

South African Bee Journal, PO Box 47198, Parklands, Johannesburg.

Speedy Bee, PO Box 998, Jesup, Ga. 31545, USA.

New Zealand Beekeeper, Box 4048, Wellington.

Book suppliers

Bee Books Old and New, Tapping Wall Farm, Burrowbridge, Bridgwater, Somerset.

Honeyfield Books, 165 Cavendish Meads, Sunninghill, Berks.

Northern Bee Books, Scout Bottom Farm, Mytholmroyd.

Suppliers

Budget Beekeeping, Kirkandrews-on-Eden, Carlisle.

Burgess & Son, Bee Equipment, 164 Fore Street, Exeter.

Exeter Bee Supplies, Unit 1D, Betton Way Industrial Estate, Moretonhampstead, Devon.

Kemlea Bee Supplies, Starcroft Apiaries, Catsfield, Battle, East Sussex.

Steele & Brodie, Beehive Works, Wormit, Fife.

B. J. Sherriff, Five Pines, Mylor Down, Falmouth, Cornwall.

E. H. Thorne Ltd, Beehive Works, Wragby, Lincoln.

USEFUL ADDRESSES – *Appendix B*

General

BBKA General Secretary, Mr Coward, 'High Trees', Dean Lane, Merstham, Surrey.

BBKA Examinations Secretary, Mr Barber, Charnwood, Beechfield, Barlaston, Stoke-on-Trent, Staffs.

BBKA Insurance, BBKA, c/o NAC Stoneleigh, Kenilworth, Warwick.

Bee Diseases Insurance, M. Wakeman, Pump Cottage, Hill Lane, Weatheroak, Alvechurch, Worcs.

Bee Farmers' Association, K. A. J. Ellis, 22 York Gardens, Clifton, Bristol.

British Isles Bee Breeders' Association, A. Knight, 11 Thomson Drive, Codnor, Derby, DE5 9RU.

Central Association of Beekeepers, Mrs K. Creighton, Long Reach, Stockbury Valley, Sittingbourne, Kent.

Devon BKA, Mr Rouse, Westholm, Braunton Road, Barnstaple, Devon.

International Bee Research Association, Hill House, Gerrards Cross, Bucks.

Ministry of Agriculture, Luddington Experimental Station, Stratford-on-Avon, Warwickshire.

National Honey Show, Mr Martin, Gander Barn, Southfields Road, Woldingham, Surrey.

Appendix C
Metric and imperial units–conversion tables

Temperature

°C	0	2	4	6	8	10	12	14	16	18	20	22	24	26	28	30	32	34	36 °C
°F	32	36	39	43	46	50	54	57	61	64	68	72	75	79	82	86	90	93	97 °F

°C	38	40	42	44	46	48	50	52	54	56	58	60	62	64	66	68	70	72	°C
°F	100	104	108	111	115	118	122	126	129	133	136	140	144	147	151	154	158	162	°F

°C	74	76	78	80	82	84	86	88	90	92	94	°C
°F	165	169	172	176	180	183	187	190	194	198	201	°F

Weights

g	7	14	28	57	85	113	142	170	198	227	255	284	312	340	368	454	g
oz	$\frac{1}{4}$	$\frac{1}{2}$	1	2	3	4	5	6	7	8	9	10	11	12	13	16	oz

kg	$\frac{1}{2}$	1	2	3	4	5	6	7	8	9	10	11	12	13	14	15	20	25	30	40	kg
lb	1.1	2.2	4.4	6.6	8.8	11	13	15	18	20	22	24	26	29	31	33	44	55	66	88	lb

Volume

litre	4.5	6.8	9	14	18	23	28	35	42	57	71	85	113	142	170	227	litre
gallon	1	$1\frac{1}{2}$	2	3	4	5	6.2	7.8	9.3	12	16	19	25	31	38	50	gallon
cub. ft	$\frac{1}{6}$	$\frac{1}{4}$	$\frac{1}{3}$	$\frac{1}{2}$	$\frac{2}{3}$	$\frac{4}{5}$	1	$1\frac{1}{4}$	$1\frac{1}{2}$	2	$2\frac{1}{2}$	3	4	5	6	8	cub. ft

Length

mm	1.3	1.6	2.5	3.2	4.8	6.4	7.9	9.5	11	13	16	19	22	25	32	mm
in	$\frac{1}{20}$	$\frac{1}{16}$	$\frac{1}{10}$	$\frac{1}{8}$	$\frac{3}{16}$	$\frac{1}{4}$	$\frac{5}{16}$	$\frac{3}{8}$	$\frac{7}{16}$	$\frac{1}{2}$	$\frac{5}{8}$	$\frac{3}{4}$	$\frac{7}{8}$	1	$1\frac{1}{4}$	in

cm	2.5	5.1	7.6	10	13	15	18	20	23	25	28	30	33	36	38	41	43	46	51 cm
in	1	2	3	4	5	6	7	8	9	10	11	12	13	14	15	16	17	18	20 in

1 oz = 28.35 g
1 lb = 453.6 g
1 ft^3 = 28.32 l
1 gallon = 4.546 l
1 in^2 = 6.452 cm^2
1 ft^2 = 929.0 cm^2
1 inch = 2.54 cm
1 ft = 30.48 cm
1 yd = 91.44 cm
1 mile = 1.609 km

Bibliography

ALFORD, D. V., *Bumblebees*, Davis-Poynter 1975.

APIMONDIA, *Hive Products*, Proceedings of International Symposium on Apitherapy, Madrid 1974.

APIMONDIA, *Propolis*, Proceedings of International Symposium on Apitherapy, Bucharest 1978.

BAILEY, Leslie, *Honey Bee Pathology*, Academic Press 1981.

BROWN, Ron, *Beeswax*, BBNO, Burrowbridge, Somerset 1981.

BROWN, Ron, *Honey Bees: A Guide to Management* 1988.

BUTLER, Colin, *The World of the Honeybee*, Collins 1954.

CRANE, Eva, *The Archaeology of Beekeeping*, Duckworth 1983.

CRANE, Eva, *Honey, a Comprehensive Survey*, Heinemann 1975.

DADANT & SONS, *The Hive and the Honeybee*, Dadant & Sons 1975.

DADE, H. A., *Anatomy and Dissection of the Honeybee*, IBRA 1962.

DIGGES, J. G., *The Practical Bee Guide*, Talbot Press 1936.

FRASER, Malcolm, *History of Beekeeping in Britain*, IBRA 1958.

FRISCH, Karl von, *The Dancing Bees*, Methuen 1955.

HARDING, Joan, *British Bee Books 1500–1976*, IBRA 1979.

HODGES, Dorothy, *The Pollen Loads of the Honey Bee*, IBRA 1952.

HOOPER, Ted, *Guide to Bees and Honey*, Blandford, Dorset 1976.

HOOPER, Ted and MORSE, Roger, *Encyclopaedia of Beekeeping* 1985.

HOWES, Frank, *Plants and Beekeeping*, Faber & Faber 1945.

HUBER, Francis, *New Observations (on Bees)*, Longman 1821.

LYALL, Neil, & CHAPMAN, Robert, *The Secret of Staying Young*, Pan Books 1976.

MANLEY, R. O. B., *Beekeeping in Britain*, Faber & Faber 1948.

MORSE, Roger, *Honey Bee Pests, Predators & Diseases*, Cornell Univ. Press 1979.

PHIPPS, John, *Beekeepers' Annual*, Northern Bee Books 1985.

SAWYER, Rex, *Pollen Identification for Beekeepers*, Cardiff Univ. Press 1981.

SIMMINS, S., *A Modern Bee Farm*, Hepworth 1888.

SNODGRASS, R. E., *Anatomy of the Honey Bee*, Cornell Univ. Press, 1956.

VERNON, Frank, *Teach Yourself Beekeeping*, Hodder and Stoughton 1976.

WEDMORE, E. B., *A Manual of Beekeeping*, Edward Arnold 1948, recently reprinted by BBNO.

Some of these books are out of print but are occasionally available second hand.

Index

Index

Index

Index

Withdrawn